W9-AOH-080

Die Grundlehren der mathematischen Wissenschaften

in Einzeldarstellungen
mit besonderer Berücksichtigung
der Anwendungsgebiete

Band 146

François Treves

Locally Convex Spaces and Linear Partial Differential Equations

Springer-Verlag Berlin Heidelberg New York 1967

Professor François Treves
Purdue University, Lafayette,
Indiana, USA.

 Library of Congress Catalog Card Number 67-25286

Printed in the Netherlands

Title-No. 5129

To Ursula

Introduction

It is hardly an exaggeration to say that, if the study of general topological vector spaces is justified at all, it is because of the needs of distribution and Linear PDE* theories (to which one may add the theory of convolution in spaces of holomorphic functions). The theorems based on TVS** theory are generally of the "foundation" type: they will often be statements of equivalence between, say, the existence – or the approximability – of solutions to an equation $\mathbf{P}u=v$, and certain more "formal" properties of the differential operator $\mathbf{P}$, for example that $\mathbf{P}$ be elliptic or hyperbolic, together with properties of the manifold X on which $\mathbf{P}$ is defined. The latter are generally geometric or topological, e.g. that X be $\mathbf{P}$-convex (Definition 20.1). Also, naturally, suitable conditions will have to be imposed upon the data, the v's, and upon the stock of possible solutions u. The effect of such theorems is to subdivide the study of an equation like $\mathbf{P}u=v$ into two quite different stages. In the first stage, we shall look for the relevant equivalences, and if none is already available in the literature, we shall try to establish them. The second stage will consist of checking if the "formal" or "geometric" conditions are satisfied. Each one of these phases requires specific techniques: checking of the formal or the geometrical conditions generally demands "hard analysis" methods, might for instance require the construction of a fundamental solution, or the proof of uniqueness in a Cauchy problem. The proof of the equivalences – the first step – will usually rely on "soft analysis", that is, on the study of rather poor structures, such as those of some brand of locally convex spaces.

The present book is concerned with the soft analysis, applied to linear PDE's. It is essentially expository, and does not contain any new result on the subject of partial differential equations. Nevertheless, it has seemed to me that there was still some room, among the publications on this subject, for a short monograph, providing the statements and the proofs of most existence and approximation theorems in the field – furthermore,

* Throughout the book, PDE will stand for "partial differential equation".
** TVS will stand for "topological vector spaces".

providing them in a general, hence flexible, form, disengaged from technicalities – as much as it could be done.

However, the usefulness of such a book, which is at any rate moderate, would not have seemed to me a sufficient reason for its writing. The determining reason lies elsewhere: namely, in the possibility of giving a new, and greatly simplifying, presentation of the basic functional analysis. Teaching of the subject before many, and widely different, audiences had convinced me that some attempt at simplifying this presentation should be made. Lately, and gradually, the ways and means of such a simplification have emerged. They go surprisingly far. But let me try to convince the reader that some improvement was needed, if only from the pedagogical point of view. A good example is provided by the following two theorems, due to S. BANACH, often applied to PDE theory:

Theorem I. – *Let E, F be two Fréchet spaces, $u: E \to F$ a continuous linear map. The mapping u is surjective* (i.e., onto) *if and only if its transpose ${}^t u: F' \to E'$ is injective and has a weakly closed image* (i.e., range).

Theorem II. – *Let E be a Fréchet space, M' a linear subspace* (or a convex subset) *of the dual E' of E. In order that M' be weakly closed, it is necessary and sufficient that the intersection of M' with any equicontinuous subset H' of E' be weakly closed in H'.*

The proof of Theorem I makes use of the open mapping theorem, of the general fact that u is a weak homomorphism of E onto its image, $\mathrm{Im}\, u \subset F$, if and only if $\mathrm{Im}\, {}^t u$ is weakly closed in E', of the special fact that, for *metrizable* locally convex spaces (here, $E/\mathrm{Ker}\, u$ and $\mathrm{Im}\, u$) weak and "strong" isomorphisms are one and the same thing. The last assertion follows essentially from the fact that, in a metrizable locally convex space, a closed convex circled subset of E is a neighborhood of O if and only if it "swallows" any weakly convergent sequence. Careful study must be made of weak topologies, especially of the weak topologies on quotient spaces.

The key lemma, in proving Theorem II, is rather tricky. It applies again to a metrizable locally convex space E. It asserts that a subset U' of the dual E' of E is open, in the sense of the topology of uniform convergence on the compact subsets of E, if and only if its intersection $U' \cap H'$ with arbitrary equicontinuous subsets H' of E' is *weakly* open. Then one exploits the following results: (i) if E is metrizable and complete, i.e., if E is a Fréchet space, the topology of uniform convergence on the compact subsets of E is identical with the topology of uniform convergence on the compact *and convex* subsets of E; (ii) when E' carries the latter, its dual is (canonically) identical with E, hence, the closure of any convex subset of E' in this topology is the same as its weak closure.*

Both proofs have, in common, the feature that they involve, at some point, three different topologies – either on the duals, or on certain quotient spaces of the duals. The propensity of TVS theory to compare and to deal with several distinct topologies at the same time, on the same underlying space, this is what nonspecialists and students resent most – not unreasonably. For it is not unreasonable to expect that a reputedly semitrivial theory be of easy access.

Our exposition, from beginning to end, allows no topology to interfere, other than the one initially given on the space E under study. The topology of E is "embodied" in the set of all the continuous seminorms on E. This set is denoted by $\mathrm{Spec}\,E$ and called *spectrum* of E – for want of a better word.** It carries no topology: its natural structure of convex cone, and the properties of the order relation between nonnegative functions, here the seminorms, will suffice for our needs. The spectrum of E contains the total information we have about E. There is no reason why we should not take full advantage of it and why, during the proofs, we should limit ourselves to dealing only with those special seminorms, the absolute values of the linear functionals. In this manner, one can go rather far, as shown in Chapters I to IV. Chapter V, which is the last one in Part I, devoted to abstract functional analysis, describes the role left to duality. For there is a role left to duality, and quite an important one! Indeed, the study carried through from Chapter I to Chapter IV leads to the two main theorems of Part I – the *epimorphism theorem* and the *theorem of existence and approximation of solutions* to a functional equation. Both theorems are concerned with a linear map $u: E \to F$ and state, under suitable hypotheses, necessary and sufficient conditions in order that the equation $u(x) = y$ be "solvable" or in order that its solutions be approximated by solutions of a simpler kind. The conditions in question bear on general (continuous) seminorms, defined either in E or in F. But when the proofs of the general theorems are completed, when we come to checking if the conditions are satisfied, then we wish to be allowed to deal with the smallest possible set of seminorms, precisely with the seminorms, absolute value of a linear functional. True, it is not only because they are fewer that these are the natural candidates to the role of "test seminorms".

* Theorem I will be proved in Section 17; the proof given will be different from the one just outlined. Theorem II will never be applied and will not be proved.

** Actually, *speculum* = mirror would have been a better interpretation of Spec.

A more compelling reason is that distribution theory and the tools of harder analysis, such as Fourier-Laplace transformation, derivation of a priori estimates, pseudodifferential operators, etc., enable us to investigate in great depth the properties of linear functionals (generally, these are distributions of certain types). While, at least for the present, we are quite disarmed when faced with general convex functionals. The last section of Part I, n° 17, gives the "duality translation" (Theorems 17.1 and 17.2) of the epimorphism theorem and of the theorem on existence and approximation of solutions. The possibility of such a translation is based on one hand on the Hahn-Banach theorem, which was to be expected, essentially on the version that says that *a seminorm is the upper enveloppe of the absolute values of the linear functionals which are at most equal to it*, on the other hand, on Mackey's theorem, stating that *a seminorm* $\mathfrak{p}$ *is bounded on a set* $B \subset E$ *if and only if every linear functional* x' *on* E, *such that* $|x'| \leqslant \mathfrak{p}$, *is bounded on the set* B.

Part II presents the most important applications of Theorems 17.1 and 17.2 to linear PDE's. In most cases, the applications require no more introduction that the definition of the differential operators and of the functions and distributions spaces on which they act. For there lies another advantage of the "theory" developed in Part I: that the abstract theorems are much closer to their concrete forms than were their more classical counterparts, like Theorems I and II above: a priori estimates, semiglobal solvability (Definition 20.2), **P**-convexity (Definition 20.1) make sense, and very naturally, in the abstract set-up (see Definitions 11.1 and 11.2). One sees thus that the theorems about partial differential equations, presented in Part II, are really results of pure, and very simple, functional analysis.

In Part I, the only result which was admitted without a proof was the Hahn-Banach theorem. In Part II, I have felt free to use, without proving them, a certain number of facts about partial differential equations (or about analytic functions and analytic functionals), referring the reader to the existing literature, mainly to the books of Hörmander [*1*] and of Treves [*1*].

Certain topics, closely related to the material discussed in Part II, have been omitted. This is true of the numerous applications of Theorem 17.2 to the theory of "differential equations of infinite order", i.e., of convolution in spaces of holomorphic functions. They are straightforward applications and duplicate, in somewhat more involved situations, what is done for a linear partial differential operator acting on entire functions

(Theorem 29.1) or on holomorphic functions in convex open subsets of the complex space $\mathbf{C}^n$ (Theorem 30.2). Another important result which will not be found here is the theorem on existence of solutions to a linear PDE with constant coefficients, in the space of *all* distributions. It is due to L. HÖRMANDER ([*1*], Theorem 3.6.4). The sufficiency part of this result is easy to establish in purely abstract terms, but I have not succeeded in finding *necessary and sufficient* conditions that generalize those given by HÖRMANDER. The knowledge of such conditions would clarify the variable coefficients case.

As it stands, Part II of the present book is hardly more than an evolved version of Chapter I of MALGRANGE's thesis [*1*]. Of course, since 1955, some gain has been made in generality; more facts are known – the general picture is clearer.

I have added, at the end of the book, a summary of the definitions and the results concerning the spectrum of a locally convex space, that is, a summary of Part I. In this summary, no proofs are given. But since the proofs are all very simple, reading of it should provide a fairly accurate idea of what it is all about. Furthermore, the reader not too familiar with PDE theory will find, also at the end of the book, a very short dictionary, listing and explaining the definitions most often used about PDE (such as elliptic equation, fundamental solution, etc.).

Paris, December 1966. F. TREVES

Table of Contents

Part I.
The Spectrum of a Locally Convex Space

Chapter I. The Spectrum of a Locally Convex Space

1. Seminorms on a vector space

Throughout the forthcoming, E will denote a vector space over a field $\mathbf{K}$ which will always be either the field of real numbers, $\mathbf{R}$, or the one of complex numbers, $\mathbf{C}$.

A *seminorm* $\mathfrak{p}$ on E is a nonnegative function on E with the following properties:

(SN_1) *For all* $x \in E$ *and all* $\lambda \in \mathbf{K}$, $\mathfrak{p}(\lambda x) = |\lambda|\, \mathfrak{p}(x)$.

(SN_2) *For all* $x, y \in E$, $\mathfrak{p}(x+y) \leqq \mathfrak{p}(x) + \mathfrak{p}(y)$.

Condition (SN_1) implies $\mathfrak{p}(0)=0$. If $\mathfrak{p}(x)=0$ implies $x=0$, one says that $\mathfrak{p}$ is a *norm*.

The kernel of $\mathfrak{p}$, $\operatorname{Ker}\mathfrak{p} = \{x \in E;\ \mathfrak{p}(x)=0\}$ is a vector subspace of E.

The *unit semiball* (or the *closed* unit semiball, if precision is needed) of $\mathfrak{p}$ will be, by definition, the set

$$B_{\mathfrak{p}} = \{x \in E;\ \mathfrak{p}(x) \leqq 1\};$$

the *open unit semiball* of $\mathfrak{p}$ will be the set

$$\mathring{B}_{\mathfrak{p}} = \{x \in E;\ \mathfrak{p}(x) < 1\}.$$

It should be noted that $B_{\mathfrak{p}}$ is no more closed that $\mathring{B}_{\mathfrak{p}}$ is open, as E carries no topology (but, of course, the adjectives can be, and will be, easily justified).

If the unit semiball of a seminorm $\mathfrak{p}$ is contained in the one of another seminorm, $\mathfrak{q}$, we have $\mathfrak{q} \leqslant \mathfrak{p}$. Therefore, if two seminorms have the same unit semiball, they are equal.

We recall a few definitions relative to a subset A of a vector space. The set A is said to be *balanced* if for all $x \in A$ and all scalars λ such that $|\lambda| \leqq 1$, we have $\lambda x \in A$. The set A is said to be *absorbent* if to every $x \in E$, there is a number $\varrho > 0$ such that $x \in \lambda A$ for all scalars λ such that $|\lambda| < \varrho$. Assuming that A is balanced, in order that A be absorbent it suffices that to every $x \in E$ there is $\varrho > 0$ such that $x \in \varrho A$.

The sets $B_{\mathfrak{p}}$ and $\mathring{B}_{\mathfrak{p}}$ are convex, balanced, absorbent.

Let A be a convex balanced absorbent subset of E. The function of $x \in E$,

$$\mathfrak{p}_A(x) = \inf_{\varrho > 0,\ x \in \varrho A} \varrho ,$$

is called the *gauge* of A. It is easily seen to be a seminorm. Distinct sets may have the same gauge. If $\mathfrak{p}$ is a seminorm, the open unit semiball of $\mathfrak{p}$ is the smallest set having $\mathfrak{p}$ as gauge, whereas its closed unit semiball is the largest one.

We shall often make use of the extension result below. Although it is much less deep (and much easier to prove), it bears some similarity to the Hahn-Banach theorem on extensions of linear functionals:

Lemma 1.1. – *Let M be a vector subspace of E, $\mathfrak{p}$ a seminorm on M, $\mathfrak{q}$ a seminorm on E whose restriction to M is $\geqslant \mathfrak{p}$. Then there exists a seminorm $\bar{\mathfrak{p}}$ on E which extends $\mathfrak{p}$ and is $\leqslant \mathfrak{q}$.*

Proof: Let $B_\mathfrak{p} \subset M$ be the unit semiball of $\mathfrak{p}$, $B_\mathfrak{q} \subset E$ the one of $\mathfrak{q}$; we have $B_\mathfrak{q} \cap M \subset B_\mathfrak{p}$. Let B be the set of points of E of the form

$$tx + (1-t)\,y, \quad x \in B_\mathfrak{p}, \quad y \in B_\mathfrak{q}, \quad 0 \leqq t \leqq 1 .$$

One checks at once that B is convex and balanced; it contains $B_\mathfrak{q}$, therefore it is absorbent. Its gauge $\bar{\mathfrak{p}}$ is $\leqslant \mathfrak{q}$. Let $z \in B \cap M$. We may write $z = tx + (1-t)\,y$ ($x \in B_\mathfrak{p}$, $y \in B_\mathfrak{q}$, $0 \leqq t \leqq 1$). If $t=1$, we have $z=x$, hence $z \in B_\mathfrak{p}$. If $t \neq 1$, $y = (1-t)^{-1}(z - tx) \in M$, hence $y \in M \cap B_\mathfrak{q}$, hence $y \in B_\mathfrak{p}$. Since $B_\mathfrak{p}$ is convex, we have $z \in B_\mathfrak{p}$. Thus we see that $B \cap M = B_\mathfrak{p}$. This implies that the restriction of $\bar{\mathfrak{p}}$ to M is equal to $\mathfrak{p}$. Q.E.D.

2. The spectrum of a locally convex space

Let E be a vector space over the field $\mathbf{K} = \mathbf{R}$ or $\mathbf{C}$.

Definition 2.1. – *A set S of seminorms on E is called irreducible if S satisfies the following conditions:*

$$S \text{ is a cone, i.e. } \mathfrak{p} \in S \text{ and } t \geqslant 0 \text{ implies } t\mathfrak{p} \in S; \tag{2.1}$$

$$\text{if } \mathfrak{p} \in S \text{ and if a seminorm } \mathfrak{q} \text{ on } E \text{ is } \leqslant \mathfrak{p}, \text{ then } \mathfrak{q} \in S; \tag{2.2}$$

$$\text{if } \mathfrak{p}_1 \text{ and } \mathfrak{p}_2 \text{ belong to } S, \text{ so does } \sup(\mathfrak{p}_1, \mathfrak{p}_2). \tag{2.3}$$

The collection of all the seminorms on E is an irreducible set. This is another way of saying that the supremum of two seminorms is a seminorm. Arbitrary intersections of irreducible sets of seminorms are irreducible. Given any set of seminorms on E, the intersection of all the irreducible sets of seminorms on E which contain it will be called the *irreducible hull* of that set.

The term "irreducible" draws some justification from the next statement:

Proposition 2.1. – *The union $S_1 \cup \ldots \cup S_m$ of a finite family of irreducible sets of seminorms on E is irreducible if and only if one of the sets S_j ($1 \leqslant j \leqslant m$) contains all the others.*

Proof: The "if" part is obvious. In order to prove the "only if" part, we reason by contradiction and assume that none of the sets S_j contains all the others. For each $j = 1, \ldots, m$, there is, in this case, an element $\mathfrak{p}_j$ of $\bigcup_{1 \leqslant k \leqslant m, k \neq j} S_k$ which does not belong to S_j. If $S_1 \cup \ldots \cup S_m$ were irreducible, $\mathfrak{p} = \sup(\mathfrak{p}_1, \ldots, \mathfrak{p}_m)$ would belong to it, hence to at least one of the sets S_k, say S_j. But then, in virtue of (2.2), $\mathfrak{p}_j \leqslant \mathfrak{p}$ should also belong to S_j and this is not true. Q.E.D.

Let f be a seminorm on a finite dimensional Euclidean space $\mathbf{R}^n$ having the following property:

Let $s = (s_1, \ldots, s_n)$, $t = (t_1, \ldots, t_n)$ be any two points of $\mathbf{R}^n$ such that

$$0 \leqslant s_j \leqslant t_j \quad \textit{for every} \quad j = 1, \ldots, n. \tag{2.4}$$

Then $f(s) \leqq f(t)$.

Proposition 2.2. – *Let S be an irreducible set of seminorms in E, f a seminorm on $\mathbf{R}^n$ having Property (2.4). Let $\mathfrak{p}_1, \ldots, \mathfrak{p}_n$ be n seminorms on E belonging to S. For every $x \in E$, set*

$$\mathfrak{p}(x) = (\mathfrak{p}_1(x), \ldots, \mathfrak{p}_n(x)) \quad (\in \mathbf{R}^n).$$

The seminorm in E, $f \circ \mathfrak{p} : x \rightsquigarrow f(\mathfrak{p}(x))$, belongs to S.

Proof: Property (2.4) insures that $f \circ \mathfrak{p}$ is indeed a seminorm on E. Now, to every seminorm f on $\mathbf{R}^n$, there is a constant $C > 0$ such that, for all $t = (t_1, \ldots, t_n)$,

$$f(t) \leqslant C \sup_{1 \leqslant j \leqslant n} |t_j|.$$

It suffices to take $C = \sum_{j=1}^{n} f(e_j)$, where $(e_1, \ldots, e_n)$ is the canonical basis of $\mathbf{R}^n$. Thus we have

$$f \circ \mathfrak{p} \leqslant C \sup(\mathfrak{p}_1, \ldots, \mathfrak{p}_n).$$

The right hand side belongs to S in view of (2.1) and (2.3), and so does therefore the left hand side, in view of (2.2).

Corollary. – *If $\mathfrak{p}_1, \ldots, \mathfrak{p}_n$ belong to an irreducible set S, the same is true of the functions*

$$(\mathfrak{p}_1^\alpha + \cdots + \mathfrak{p}_n^\alpha)^{1/\alpha}, \quad 1 \leqslant \alpha < +\infty.$$

Definition 2.2. – *The pair (E, S) consisting of a vector space E and of an irreducible set S of seminorms on E is called a locally convex space. The set S is called the spectrum of the locally convex space.*

We shall write E rather than (E, S) and set $S = \operatorname{Spec} E$. If E is a locally convex space, we say that a subset U of E is a neighborhood of O if it contains the unit semiball of some seminorm $\mathfrak{p} \in \operatorname{Spec} E$; a subset V of E will be a neighborhood of a point x of E if $V = x + U$, U being a neighborhood of O. This defines a topology on E having the properties that $(x, y) \rightsquigarrow x + y$ is a continuous map of $E \times E$ (equipped with the product topology) into E and that $(x, y) \rightsquigarrow \lambda x$ is a continuous map of $\mathbf{K} \times E$ into E. Thus, in a locally convex space, we may talk of open and closed sets, of convergent filters (or nets), sequences, etc.

Sometimes it will be convenient to deal with the set of all the seminorms on a vector space E. We shall refer to it as the *total spectrum* of E and denote it by $\operatorname{SPEC} E$. If then E carries a locally convex topology, its spectrum, $\operatorname{Spec} E$, is an irreducible subset of $\operatorname{SPEC} E$. The locally convex topologies on E form a partially ordered set, for the obvious natural order relation: inclusion of irreducible subsets of $\operatorname{SPEC} E$. As a matter of fact, this set is a lattice: the *infimum* of two topologies, defined by irreducible subsets of $\operatorname{SPEC} E$, S_1 and S_2, respectively, is the topology defined by the intersection $S_1 \cap S_2$; their supremum is defined by the irreducible hull of $S_1 \cup S_2$. Obviously, this lattice has a maximum element: the finest locally convex topology on E, for which the spectrum of E is identical with its total spectrum, $\operatorname{SPEC} E$, and a minimum element, the trivial topology on E, for which the spectrum reduces to the seminorm identically equal to zero.

Proposition 2.3. – *The following properties of a seminorm $\mathfrak{p}$ on a locally convex space E are equivalent:*

(a) $\mathfrak{p} \in \operatorname{Spec} E$;

(b) $\mathfrak{p}$ *is uniformly continuous in* E;

(c) $\mathfrak{p}$ *is continuous in* E;

(d) $\mathfrak{p}$ *is continuous at the origin*;

(e) *the open unit semiball of* $\mathfrak{p}$, $\mathring{B}_{\mathfrak{p}}$, *is an open subset of* E.

Proof: The implications (b)⇒(c)⇒(d)⇒(e) are trivial. Suppose that $\mathfrak{p} \in \operatorname{Spec} E$. Let $\varepsilon > 0$ arbitrary, $x, y \in E$ such that

$$x - y \in \varepsilon B_{\mathfrak{p}} = B_{(1/\varepsilon)\mathfrak{p}}\,.$$

Then

$$|\mathfrak{p}(x) - \mathfrak{p}(y)| \leqslant \mathfrak{p}(x - y) \leqslant \varepsilon;$$

in view of (2.1), $(1/\varepsilon)\,\mathfrak{p}$ belongs to Spec$E$, whence (b). Lastly, suppose that (e) holds: if $\mathring{B}_{\mathfrak{p}}$ is open, it contains some set $B_{\mathfrak{q}}$ with $\mathfrak{q}\in\operatorname{Spec}E$ (since then $\mathring{B}_{\mathfrak{p}}$ is a neighborhood of O), hence $\mathfrak{p}\leqslant\mathfrak{q}$. In virtue of (2.2), $\mathfrak{p}\in\operatorname{Spec}E$.

Q.E.D.

Thus Spec$E$ consists exactly of the continuous seminorms on E.

Definition 2.3. – *A subset Σ of* Spec$E$ *is called a basis of continuous seminorms on E if to every $\mathfrak{q}\in\operatorname{Spec}E$ there is $\mathfrak{p}\in\Sigma$ and a number $C>0$ such that $\mathfrak{q}\leqslant C\mathfrak{p}$.*

One says often that a subset Σ' of Spec$E$ *defines the topology* of E if the irreducible hull of Σ' is equal to Spec$E$. This is equivalent with the fact that the seminorms $\sup_{\mathfrak{p}\in S}\mathfrak{p}$, as S runs over the family of finite subsets of Σ', form a basis of continuous seminorms on E.

3. Polar. Bipolar

We shall need the following "separation" result:

Lemma 3.1. – *Let E be a locally convex space, A a closed convex balanced subset of E, K a compact subset of E which does not intersect A.*

There is a continuous seminorm $\mathfrak{p}$ in E such that $\mathfrak{p}\leqslant 1$ in A and $\mathfrak{p}>1$ in K.

Proof: For every neighborhood of O, V, in E, let Ω_V be the complement of the closure of $A+V$. Let x be an arbitrary point of K. There is a neighborhood of O, U, such that $x+U$ does not intersect A. Let V be a balanced neighborhood of O such that $V+V\subset U$; then $x+V$ does not intersect $A+V$. If we take V open, we may even assert that $x+V$ does not intersect $\overline{A+V}$, i.e. $x+V\subset\Omega_V$. This shows that the open sets Ω_V form a covering of K. From the compactness of K follows then that K must be contained in the union of a finite family of Ω_V; but such a family is contained in some Ω_V, say Ω_U. Consider another neighborhood of O, U', such that $U'+U'\subset U$; moreover, take U' to be open convex balanced. Then $A+U'$ and $K+U'$ do not intersect otherwise $A+U$ and K would. Since $K+U'$ is open, $\overline{A+U'}$ does not intersect $K+U'$. But $\overline{A+U'}$ is closed convex balanced; as it contains U', it is a neighborhood of O. Its gauge $\mathfrak{p}$ satisfies the requirements of Lemma 3.1.

Let us underline the fact that the compact set K, in Lemma 3.1, is arbitrary: it need not be convex!

Let M be an arbitrary subset of the locally convex space E. The *closed convex balanced hull* of M is the smallest closed convex balanced subset of E containing M.

Definition 3.1. – *We call polar of M the set of seminorms $\mathfrak{p} \in \operatorname{Spec} E$ such that $\mathfrak{p}(x) \leqslant 1$ for all $x \in M$.*

The polar of M will be denoted by M^0; it possesses Properties (2.2) and (2.3) (Definition 2.1) but not, at least in general, (2.1). Suppose that M is a cone, i.e., that $tx \in M$ for all $x \in M$ and all numbers $t \geqslant 0$. Then, if $\mathfrak{p}(x) \leqslant 1$ for all $x \in M$, we have also $\mathfrak{p}(tx) = t\mathfrak{p}(x) \leqslant 1$ for all $x \in M$ and all numbers $t > 0$, which implies $\mathfrak{p}(x) = 0$. Thus, when M is a cone,

$$M^0 = \{\mathfrak{p} \in \operatorname{Spec} E;\ \text{for all } x \in M,\ \mathfrak{p}(x) = 0\}.$$

Note, then, that M^0 is an irreducible set.

Let now Σ be a subset of $\operatorname{Spec} E$.

Definition 3.2. – *We call kernel of Σ the set of points $x \in E$ such that $\mathfrak{p}(x) = 0$ for all $\mathfrak{p} \in \Sigma$. We call polar of Σ the set of points $x \in E$ such that $\mathfrak{p}(x) \leqslant 1$ for all $\mathfrak{p} \in \Sigma$.*

The kernel of Σ will be denoted by $\operatorname{Ker} \Sigma$, its polar by Σ^0. The two notions coincide when Σ is a cone. In any case, $\operatorname{Ker} \Sigma$ is a closed vector subspace of E; Σ^0 is a closed convex balanced set. The kernel of the full $\operatorname{Spec} E$ may be strictly larger than $\{0\}$. As a matter of fact, we have

$$\operatorname{Ker} \operatorname{Spec} E = \overline{\{0\}},$$

the intersection of all the neighborhoods of 0. The locally convex space E is Hausdorff if and only if $\operatorname{Ker} \operatorname{Spec} E = \{0\}$.

Going to the polars (either in E or in $\operatorname{Spec} E$) reverses the inclusion relation:

$$A \subset B \quad \text{implies} \quad B^0 \subset A^0. \tag{3.1}$$

Also, if ϱ is a number >0,

$$(\varrho A)^0 = \varrho^{-1} A^0. \tag{3.2}$$

The next identity is also obvious:

$$(A \cup B)^0 = A^0 \cap B^0. \tag{3.3}$$

Definition 3.3. – *Let M be a subset of E, M^0 its polar. The polar M^{00} of M^0 is called the bipolar of M.*

Proposition 3.1. – *The bipolar M^{00} of M is the closed convex balanced hull of M.*

Proof: Let $\hat{M}$ be the closed convex balanced hull of M. If a continuous seminorm $\mathfrak{p}$ is $\leqslant 1$ on M, it is also $\leqslant 1$ on $\hat{M}$. Hence $\hat{M} \subset M^{00}$. Suppose there were a point $x \in M^{00}$, $x \notin \hat{M}$. Lemma 3.1 implies the existence of a continuous seminorm $\mathfrak{p}$ on E such that $\mathfrak{p} \leqslant 1$ on $\hat{M}$ (hence on M) and $\mathfrak{p}(x) > 1$. But this contradicts the fact that $x \in M^{00}$. Q.E.D.

Corollary. – $(M^{00})^0 = M^0$.

4. Continuous linear mappings

Let E, F be two vector spaces over the field $\mathbf{K}$, $u: E \to F$ a linear map. We consider the commutative diagram:

$$\begin{array}{ccc} E & \xrightarrow{\underline{u}} \operatorname{Im} u & \xrightarrow{j} F \\ \varphi \downarrow & \nearrow \bar{u} & \\ E/\operatorname{Ker} u\,, & & \end{array} \tag{4.1}$$

where j is the natural injection of $\operatorname{Im} u$ into F, φ the canonical surjection of E onto its quotient space $E/\operatorname{Ker} u$. The map $\underline{u}$ is uniquely determined by the fact that $u = j \circ \underline{u}$, $\bar{u}$ by the fact that Diagram (4.1) is commutative; $\underline{u}$ is *surjective* (i.e., onto), $\bar{u}$ *injective* (i.e., one-to-one) and surjective, hence *bijective*.

Suppose now that E and F are topological vector spaces (say, locally convex spaces). The quotient topology on $E/\operatorname{Ker} u$ is defined by the fact that a subset of $E/\operatorname{Ker} u$ is open if and only if it is the image of an open subset of E under φ. As for $\operatorname{Im} u$, it carries always the topology induced by F. Then we see at once that u is continuous if and only if $\bar{u}$ is continuous (and, trivially, if and only if $\underline{u}$ is).

We introduce now a terminology which shall be systematically used throughout the sequel. We assume that u is continuous. We say that u is a *homomorphism* if $\underline{u}$ is an open mapping (i.e., transforms open sets into open sets) or, which amounts to the same, if the inverse of $\bar{u}$ is continuous. A homomorphism u is called

a *monomorphism* if u is injective;

an *epimorphism* if u is surjective;

an *isomorphism* if u is bijective.

Whatever be the continuous linear map u, φ is an epimorphism (and j a monomorphism); that u be a homomorphism means that $\underline{u}$ is an epimorphism (resp. that $\bar{u}$ is an isomorphism), etc. etc.

Whatever be the seminorm $\mathfrak{q}$ on F, $x \rightsquigarrow \mathfrak{q}(u(x))$ is a seminorm on E, namely the compose $\mathfrak{q} \circ u$; if $\mathfrak{q}_1 \leqslant \mathfrak{q}_2$, we have obviously $\mathfrak{q}_1 \circ u \leqslant \mathfrak{q}_2 \circ u$. Also

$$\sup(u \circ \mathfrak{q}_1, u \circ \mathfrak{q}_2) = u \circ \sup(\mathfrak{q}_1, \mathfrak{q}_2)\,. \tag{4.2}$$

More generally, if f is a seminorm on the Euclidean space $\mathbf{R}^n$, having

Property (2.4), and if $\mathfrak{q}_1, \ldots, \mathfrak{q}_n$ are seminorms on F, we have

$$f(u \circ \mathfrak{q}_1, \ldots, u \circ \mathfrak{q}_n) = u \circ f(\mathfrak{q}_1, \ldots, \mathfrak{q}_n). \tag{4.3}$$

From now on, we suppose that E and F are locally convex spaces. It is seen at once that the mapping u is continuous if and only if, for every $\mathfrak{q} \in \operatorname{Spec} F$, we have $u \circ \mathfrak{q} \in \operatorname{Spec} E$. In this case,

$$\mathfrak{q} \rightsquigarrow u \circ \mathfrak{q}$$

is an order preserving mapping of $\operatorname{Spec} F$ into $\operatorname{Spec} E$, which we shall denote by u_*.

We are going to use the notation introduced in Section 3, Definitions 3.1 and 3.2. Setting

$$\operatorname{Ker} u_* = \{\mathfrak{q} \in \operatorname{Spec} F;\ u_* \mathfrak{q} = 0\},$$

we see at once that

$$\operatorname{Ker} u_* = (\operatorname{Im} u)^0. \tag{4.4}$$

From this we derive:

Proposition 4.1. – *The polar* $(\operatorname{Ker} u_*)^0$ *is the closure of* $\operatorname{Im} u$ *in* F.

Combine (4.4) with Proposition 3.1.

Proposition 4.2. – *The following properties are equivalent*:

(a) $\operatorname{Im} u$ *is dense*;

(b) u_* *is injective.*

Proof: If u_* is injective, $\operatorname{Ker} u_* = \{0\}$, hence $\overline{\operatorname{Im} u} = F$ in view of Proposition 4.1. Conversely, if $\operatorname{Im} u$ is dense, any two continuous seminorms on F which are equal on $\operatorname{Im} u$ must be equal everywhere.

We examine, next, the relationship between the spectrum of E and the spectra of the linear subspaces and the quotient spaces of E. The situation of a linear subspace is essentially described by the extension lemma 1.1:

Proposition 4.3. – *If* u *is a monomorphism,* u_* *is surjective.*

Proof: If u is a monomorphism and $\mathfrak{p} \in \operatorname{Spec} E$, $u(x) \rightsquigarrow \mathfrak{p}(x)$ is a continuous seminorm on $\operatorname{Im} u$, which can be extended as a continuous seminorm $\mathfrak{q}$ on F (Lemma 1.1); we have obviously $\mathfrak{p} = u_* \mathfrak{q}$.

Corollary 1. – *Let* M *be a vector subspace of* E, *equipped with the induced topology. Let* j *be the natural injection* $M \to E$. *Then* j_* *is a surjection of* $\operatorname{Spec} E$ *onto* $\operatorname{Spec} M$, *whose kernel is equal to* M^0.

Combine Proposition 4.3 with Formula (4.4).

Corollary 2. – *Let* $u: E \to F$ *be a monomorphism with dense image. Then* u_* *is a bijection of* $\operatorname{Spec} F$ *onto* $\operatorname{Spec} E$.

Combine Propositions 4.2 & 4.3.

Corollary 3. – *Let E be Hausdorff, $\hat{E}$ be its completion;* Spec E *and* Spec $\hat{E}$ *are canonically isomorphic.*

Let $J: E \to \hat{E}$ be the canonical injection; J is a monomorphism with dense image. By corollary 2, we see that J_* is a bijection of Spec $\hat{E}$ onto Spec E.

We proceed, now, with the study of the quotient space E/M (M: vector subspace of E). Let us denote by φ the canonical epimorphism $E \to E/M$. We set, for arbitrary $\mathfrak{p} \in \operatorname{Spec} E$ and $x \in E$,

$$\varphi^* \mathfrak{p}(\varphi(x)) = \sup_{y \in E,\ x - y \in M} \mathfrak{p}(y); \tag{4.5}$$

$\varphi^* \mathfrak{p}$ is a continuous seminorm on E/M. If $\mathfrak{q} \in \operatorname{Spec} E/M$, we check at once that

$$\varphi^* \varphi_* \mathfrak{q} = \mathfrak{q}. \tag{4.6}$$

Of course, we have $\varphi_* \mathfrak{q} \in M^0$ (i.e., $\varphi_* \mathfrak{q}$ vanishes on M). Conversely, let $\mathfrak{p} \in M^0$. Then, if $x - y \in M$, we have $\mathfrak{p}(x) = \mathfrak{p}(y)$, hence

$$\varphi^* \mathfrak{p}(\varphi(x)) = \mathfrak{p}(x),$$

that is,

$$\varphi_* \varphi^* \mathfrak{p} = \mathfrak{p}. \tag{4.7}$$

We may summarize this as follows:

Proposition 4.4. – *Let M be a vector subspace of a locally convex space E, φ the canonical epimorphism $E \to E/M$.*

The mapping φ_*: Spec $E/M \to$ Spec E *is an injection, and*

$$\operatorname{Im} \varphi_* = M^0.$$

The mapping φ^*: Spec $E \to$ Spec E/M, *defined in* (4.5), *is a left inverse of φ_*. Its restriction to M^0 is an inverse of φ_*; the compose $\varphi_* \varphi^*$ is a retraction of* Spec E *onto M^0.*

That φ_* is injective follows from Proposition 4.2. The left inverse property corresponds to (4.6). We have used the word "retraction" to indicate a mapping of a set onto one of its subsets which is the identity on the subset; the retraction property corresponds to (4.7). The mappings φ_* and φ^* are order preserving.

Corollary 1. – *Let $\overline{\{0\}}$ be the closure of the origin in E, π the canonical map of E onto $E/\overline{\{0\}}$, the Hausdorff TVS associated with E. Then π_* is a bijection of* Spec$(E/\overline{\{0\}})$ *onto* Spec E.

Apply Proposition 4.4, observing that, when $M = \overline{\{0\}}$, $M^0 = \operatorname{Spec} E$.

We may combine the statements of Propositions 4.3 and 4.4:

Proposition 4.5. – *Consider an exact sequence*:

$$0 \to E \xrightarrow{j} F \xrightarrow{\phi} G \to 0,$$

where j is a monomorphism and φ an epimorphism. We have then the exact sequence:

$$0 \to \operatorname{Spec} G \overset{\varphi_*}{\to} \operatorname{Spec} F \overset{j_*}{\to} \operatorname{Spec} E \to 0;$$

and $\operatorname{Im} \varphi_* = \operatorname{Ker} j_* = (\operatorname{Im} j)^0 = (\operatorname{Ker} \varphi)^0$.

We go back to the continuous linear map $u: E \to F$.

Since u is continuous, we have

$$u(\overline{\{0\}}) \subset \overline{\{0\}}, \quad \text{i.e.,} \quad \overline{\{0\}} \subset \overset{-1}{u}(\overline{\{0\}}).$$

Thus u induces a continuous linear map u_1 of $E/\overline{\{0\}}$ into $F/\overline{\{0\}}$, such that the following diagram be commutative:

$$\begin{array}{ccc} E & \overset{u}{\to} & F \\ \pi \downarrow & & \downarrow \varpi \\ E/\overline{\{0\}} & \underset{u_1}{\to} & F/\overline{\{0\}}, \end{array}$$

where π and ϖ are the canonical mappings. We have:

$$u_* = \varpi_* \circ u_{1*} \circ \pi_* . \tag{4.8}$$

In virtue of Corollary 1 of Proposition 4.4, π_* and ϖ_* are bijections. Note also that

$$\overset{-1}{u}(\overline{\{0\}}) = (\operatorname{Im} u_*)^0 . \tag{4.9}$$

In particular,

$$\text{if } F \text{ is Hausdorff,} \quad \operatorname{Ker} u = (\operatorname{Im} u_*)^0 . \tag{4.10}$$

Proposition 4.3 can now be made more precise:

Proposition 4.6. – *The following properties are equivalent*:

(a) *u induces a monomorphism $E/\overline{\{0\}} \to F/\overline{\{0\}}$*;

(b) *u_* is surjective.*

Proof: The mapping $E/\overline{\{0\}} \to F/\overline{\{0\}}$ induced by u is the mapping u_1 above. In view of (4.8), (b) is equivalent with the fact that u_{1*} is surjective. Then Proposition 4.3 shows that (a) implies (b). And in order to prove that (b) implies (a), we may assume that E and F are Hausdorff.

Suppose that (b) holds; F being now Hausdorff, (4.10) implies that u is injective. Moreover, to every $\mathfrak{p} \in \operatorname{Spec} E$ there is $\mathfrak{q} \in \operatorname{Spec} F$ such that, for all $x \in E$, $\mathfrak{p}(x) = \mathfrak{q}(u(x))$. The image under u of the unit semiball of $\mathfrak{p}$ is equal to the intersection of $\operatorname{Im} u$ with the unit semiball of $\mathfrak{q}$; thus u is an open mapping. Q.E.D.

Corollary 1. – *Let E and F be Hausdorff. The following properties are equivalent*:

(a) *u is a monomorphism;*
(b) *u_* is surjective.*

Corollary 2. – *Let E and F be Hausdorff. The following properties are equivalent:*
(a) *u is a monomorphism with dense image;*
(b) *u_* is bijective.*

Combine Corollary 1 with Proposition 4.2.

Corollary 3. – *Let E and F be Hausdorff and complete. The following conditions are equivalent:*
(a) *u is an isomorphism;*
(b) *u_* is bijective.*

Proof: (a) implies (b), by Corollary 2; by the same, if (b) holds, $\operatorname{Im} u$ is dense and $\overset{-1}{u}:\operatorname{Im} u \to E$ is a monomorphism which can be extended, as E is complete, into an isomorphism v of F into E; but, obviously, v is the inverse of u.

Let us observe now that we have always

$$\operatorname{Im} u_* \subset (\overset{-1}{u}(\overline{\{0\}}))^0, \tag{4.11}$$

in particular:

$$\textit{if } F \textit{ is Hausdorff,} \quad \operatorname{Im} u_* \subset (\operatorname{Ker} u)^0. \tag{4.12}$$

But in general, these inclusions cannot be replaced by equalities. We are now going to show what is the meaning of the corresponding equalities. For this, let us consider the commutative diagram (cf. diagram in p. 9):

$$\begin{array}{ccccc} E & \overset{\underline{u}}{\to} & \operatorname{Im} u & \overset{j}{\to} & F \\ \pi \downarrow & & \omega \downarrow & & \downarrow \varpi \\ E/\overline{\{0\}} & \overset{\underline{u}_1}{\to} & \operatorname{Im} u/\overline{\{0\}} & \overset{j_1}{\to} & F/\overline{\{0\}} \\ \phi_1 \downarrow & \nearrow \bar{u}_1 & & & \\ (E/\overline{\{0\}})/\operatorname{Ker} u_1, & & & & \end{array} \tag{4.13}$$

where j is the natural injection, π, ω, ϖ and φ the canonical epimorphisms, $\underline{u}_1$, j_1, $\bar{u}_1$ the (uniquely determined) mappings such that the diagram be commutative. We have $u = j \circ \underline{u}$.

Proposition 4.7. – *The following conditions are equivalent:*
(a) *u is a homomorphism;*
(b) $\operatorname{Im} u_* = (\overset{-1}{u}(\overline{\{0\}}))^0$.

Proof: We reason on Diagram (4.13). First of all, it is obvious that u is a homomorphism if and only if $u_1 = j_1 \circ \underline{u}_1$ is one, i.e., if and only if $\underline{u}_1$

is an epimorphism. In virtue of Proposition 4.3, j_* is surjective, hence $\mathrm{Im}\, u_* = \mathrm{Im}\underline{u}_*$. Since π_* and ω_* are bijective, we have $\mathrm{Im}\, u_* = \pi_*(\mathrm{Im}\underline{u}_{1*})$. But, as one checks easily,

$$\pi_*((\mathrm{Ker}\, \underline{u}_1)^0) = (\overset{-1}{u}(\{0\}))^0 .$$

By taking again advantage of the bijectivity of π_*, we see that (b) is equivalent with the fact that $\mathrm{Im}\, \underline{u}_{1*} = (\mathrm{Ker}\underline{u}_1)^0$, which is to say, the same fact in the case where F is Hausdorff. Finally, we see that it suffices to give the proof when u is surjective and E and F are Hausdorff, which we are going to assume from now on.

We go back to Diagram (4.1) where $\mathrm{Im}\, u = F$; we derive from it the commutative diagram

$$\begin{array}{ccc} \mathrm{Spec}\, F & \xrightarrow{u_*} & \mathrm{Spec}\, E \\ & {\scriptstyle \bar{u}_*}\searrow & \uparrow {\scriptstyle \varphi_*} \\ & & \mathrm{Spec}(E/\mathrm{Ker}\, u). \end{array}$$

In view of Proposition 4.4, $\mathrm{Im}\, \varphi_* = (\mathrm{Ker}\, u)^0 \supset \mathrm{Im}\, u_*$. Since φ_* is injective (Proposition 4.2), to say that $\mathrm{Im}\, u_* = (\mathrm{Ker}\, u)^0$ is equivalent with saying that $\bar{u}_*$ is surjective. As now F is Hausdorff, $\mathrm{Ker}\, u$ is closed in E and therefore $E/\mathrm{Ker}\, u$ is Hausdorff. In view of Corollary 1 of Proposition 4.6, the fact that $\mathrm{Im}\, u_* = (\mathrm{Ker}\, u)^0$, i.e., that $\bar{u}_*$ is surjective, is equivalent with the fact that $\bar{u}$ is a monomorphism, i.e., that u is an epimorphism (here, u is assumed to be surjective). Q.E.D.

Corollary 1. – *Let $u: E \to F$ be a continuous linear map. Suppose that F is Hausdorff. The following properties are equivalent:*

(a) *u is a homomorphism;*

(b) $\mathrm{Im}\, u_* = (\mathrm{Ker}\, u)^0$.

Corollary 2. – *Let E, F be two locally convex spaces, $u: E \to F$ a continuous linear map. The following conditions are equivalent:*

(a) *u is a homomorphism with dense image;*

(b) *u_* is injective and* $\mathrm{Im}\, u_* = (\overset{-1}{u}(\overline{\{0\}}))^0$.

Combine Propositions 4.2 and 4.7.

Corollary 3. – *Let E, F, u be as in Corollary 2; let, moreover, F be Hausdorff. The following conditions are equivalent:*

(a) *u is a homomorphism with dense image;*

(b) *u_* is injective and* $\mathrm{Im}\, u_* = (\mathrm{Ker}\, u)^0$.

Chapter II. The Natural Fibration over the Spectrum

5. The natural fibration over the spectrum

Let E be a vector space, $\mathfrak{p}$ a seminorm on E. We shall denote by $E_{(\mathfrak{p})}$ the "seminormed" space $(E, \mathfrak{p})$; this space carries the topology defined by the single seminorm $\mathfrak{p}$. The space $E_{(\mathfrak{p})}$ is Hausdorff if and only if $\mathfrak{p}$ is a norm. In any case, we denote by $E_{\mathfrak{p}}$ the associated normed space, that is, the quotient space

$$E_{(\mathfrak{p})}/\operatorname{Ker}\mathfrak{p},$$

with the norm

$$\|\dot{x}\|_{\mathfrak{p}} = \mathfrak{p}(x),$$

where $x \in E$ is any representative of the class mod $\operatorname{Ker}\mathfrak{p}$, $\dot{x}$.

Definition 5.1. – *We denote by $\hat{E}_{\mathfrak{p}}$ the Banach space obtained by completion of the normed space $E_{\mathfrak{p}}$.*

The norm in $\hat{E}_{\mathfrak{p}}$ will be denoted

$$\hat{x} \rightsquigarrow \|\hat{x}\|_{\mathfrak{p}}.$$

By composing the canonical surjection of E onto $E/\operatorname{Ker}\mathfrak{p}$ with the natural injection of $E/\operatorname{Ker}\mathfrak{p}$ into $\hat{E}_{\mathfrak{p}}$, we obtain what we shall call *the canonical map of E into $\hat{E}_{\mathfrak{p}}$* and denote by $\omega_{\mathfrak{p}}$.

Let now $\mathfrak{q}$ be another seminorm on E such that $\mathfrak{q} \leqslant \mathfrak{p}$. We have $\operatorname{Ker}\mathfrak{p} \subset \operatorname{Ker}\mathfrak{q}$, and we may transfer the identity mapping of E to the quotient spaces; we obtain thus a mapping of $E_{\mathfrak{p}}$ into $E_{\mathfrak{q}}$ which is continuous and, as a matter of fact, of norm $\leqslant 1$. We may extend, by continuity, this mapping of $E_{\mathfrak{p}}$ into $E_{\mathfrak{q}}$ as a continuous linear map (with norm $\leqslant 1$),

$$\omega_{\mathfrak{q}}^{\mathfrak{p}} : \hat{E}_{\mathfrak{p}} \to \hat{E}_{\mathfrak{q}}.$$

We have the following consistency conditions:

$$\text{If } \mathfrak{r} \leqslant \mathfrak{q} \leqslant \mathfrak{p}, \quad \omega_{\mathfrak{r}}^{\mathfrak{p}} = \omega_{\mathfrak{r}}^{\mathfrak{q}} \circ \omega_{\mathfrak{q}}^{\mathfrak{p}}. \tag{5.1}$$

$$\text{If } \mathfrak{q} \leqslant \mathfrak{p}, \quad \omega_{\mathfrak{q}} = \omega_{\mathfrak{q}}^{\mathfrak{p}} \circ \omega_{\mathfrak{p}}; \tag{5.2}$$

of course, $\omega_{\mathfrak{p}}^{\mathfrak{p}}$ is the identity mapping of $\hat{E}_{\mathfrak{p}}$.

Let now F be a second vector space (over the field $\mathbf{K}$), $u: E \to F$ a linear map. Let $\mathfrak{p}$ be a seminorm on E, $\mathfrak{q}$ one on F such that $\mathfrak{q} \circ u \leqslant \mathfrak{p}$. Then u, regarded as a linear map $E_{(\mathfrak{p})} \to F_{(\mathfrak{q})}$, is continuous; by transfer to the quotient spaces, it defines a continuous linear map of $E_\mathfrak{p}$ into $F_\mathfrak{q}$, then, extended to the completions, it defines a continuous linear map

$$u_\mathfrak{q}^\mathfrak{p}: \hat{E}_\mathfrak{p} \to \hat{F}_\mathfrak{q}.$$

Because $\mathfrak{q} \circ u \leqslant \mathfrak{p}$, the norm of $u_\mathfrak{q}^\mathfrak{p}$ is $\leqslant 1$. When $\mathfrak{p} = \mathfrak{q} \circ u$, we shall write $u_\mathfrak{q}$ rather than $u_\mathfrak{q}^{u_* \mathfrak{q}}$; $u_\mathfrak{q}$ is an *isometry* of $\hat{E}_{u_* \mathfrak{q}}$ into $\hat{F}_\mathfrak{q}$.

Let $\mathfrak{q}$ and $\mathfrak{q}'$ be two seminorms on F such that $\mathfrak{q}' \leqslant \mathfrak{q}$; we denote by $\varpi_\mathfrak{q}$ (resp. $\varpi_{\mathfrak{q}'}^\mathfrak{q}$) the canonical map $F \to \hat{F}_\mathfrak{q}$ (resp. $\hat{F}_\mathfrak{q} \to \hat{F}_{\mathfrak{q}'}$). The mappings $\varpi_\mathfrak{q}$ and $\varpi_{\mathfrak{q}'}^\mathfrak{q}$ are the analogs of $\omega_\mathfrak{p}$ and $\omega_{\mathfrak{p}'}^\mathfrak{p}$ when F is substituted for E.

Proposition 5.1. – *Let E, F be two vector spaces over* $\mathbf{K}$, *$u: E \to F$ a linear mapping. Let $\mathfrak{p}$, $\mathfrak{p}'$ be two seminorms on E, $\mathfrak{q}$, $\mathfrak{q}'$ two seminorms on F, such that*

$$\mathfrak{p} \geqslant \mathfrak{p}', \quad \mathfrak{q} \geqslant \mathfrak{q}', \quad \mathfrak{q} \circ u \leqslant \mathfrak{p}, \quad \mathfrak{q}' \circ u \leqslant \mathfrak{p}'.$$

Then, the following diagram is commutative:

$$\begin{array}{ccccc} E & \xrightarrow{\omega_\mathfrak{p}} & \hat{E}_\mathfrak{p} & \xrightarrow{\omega_{\mathfrak{p}'}^\mathfrak{p}} & \hat{E}_{\mathfrak{p}'} \\ {\scriptstyle u}\downarrow & & \downarrow{\scriptstyle u_\mathfrak{q}^\mathfrak{p}} & & \downarrow{\scriptstyle u_{\mathfrak{q}'}^{\mathfrak{p}'}} \\ F & \xrightarrow[\varpi_\mathfrak{q}]{} & \hat{F}_\mathfrak{q} & \xrightarrow[\varpi_{\mathfrak{q}'}^\mathfrak{q}]{} & \hat{F}_{\mathfrak{q}'} \end{array} \tag{5.3}$$

The proof is a matter of routine checking.

From now on, E will be a locally convex space and we focus our attention to the seminorms on E which are continuous. We shall view the disjoint union $\mathfrak{E}$ of the Banach spaces $\hat{E}_\mathfrak{p}$, as $\mathfrak{p}$ runs over $\operatorname{Spec} E$, as a *fibre set* over the base $\operatorname{Spec} E$. The fibre at $\mathfrak{p}$ is $\hat{E}_\mathfrak{p}$ and, as usual, a *section* s over a subset S of $\operatorname{Spec} E$ is a mapping of S into $\mathfrak{E}$ such that $s(\mathfrak{p}) \in \hat{E}_\mathfrak{p}$ for every $\mathfrak{p} \in S$. Let us underline the fact that no topology has been put on $\mathfrak{E}$ (nor on $\operatorname{Spec} E$).

Definition 5.2. – *We say that a section s of $\mathfrak{E}$ over a subset S of* $\operatorname{Spec} E$ *is regular if, for every couple of elements $\mathfrak{p}$, $\mathfrak{q}$ of S such that $\mathfrak{p} \geqslant \mathfrak{q}$, we have*

$$s(\mathfrak{q}) = \omega_\mathfrak{q}^\mathfrak{p}(s(\mathfrak{p})).$$

The set of regular sections over $S \subset \operatorname{Spec} E$ will be denoted by $\Gamma(S)$. Since each fibre of $\mathfrak{E}$ is a vector space, $\Gamma(S)$ carries a natural vector space structure (over the field $\mathbf{K}$). It also carries a natural topology (compatible with its vector space structure), namely *the topology of pointwise convergence in S*. This turns $\Gamma(S)$ into a Hausdorff locally convex space. Indeed, the topology of pointwise convergence on $\Gamma(S)$ is the topology defined by the seminorms

$$s \rightsquigarrow \sup_{\mathfrak{p} \in A} \|s(\mathfrak{p})\|_{\mathfrak{p}},$$

as A runs over the family of all finite subsets of S. An equivalent definition of the locally convex space $\Gamma(S)$ makes it the projective limit of the Banach spaces $\hat{E}_{\mathfrak{p}}$ with respect to the mappings $\omega_{\mathfrak{q}}^{\mathfrak{p}}$. *The space* $\Gamma(S)$ *is complete.* Indeed, a filter (or a net) of sections converges in $\Gamma(S)$ if and only if the filter of values at each point $\mathfrak{p}$ of S converges in $\hat{E}_{\mathfrak{p}}$; and it is a Cauchy filter if and only if the filter of values at each point $\mathfrak{p} \in S$ is a Cauchy filter in $\hat{E}_{\mathfrak{p}}$. As the $\hat{E}_{\mathfrak{p}}$ are Banach spaces, in particular are complete, our assertion follows at once.

Let x be an arbitrary element of E, S a subset of $\operatorname{Spec} E$ as before. Then

$$\mathfrak{p} \rightsquigarrow \omega_{\mathfrak{p}}(x)$$

is a regular section over S, which we shall denote by $\omega_S x$. The definition of the topologies in E and $\Gamma(S)$ shows at once that

$$x \rightsquigarrow \omega_S x$$

is a continuous linear map of E into $\Gamma(S)$. We shall refer to it as the canonical map of E into $\Gamma(S)$. Note that $\operatorname{Ker} \omega_S = \operatorname{Ker} S$ (Definition 3.2); in particular, $\omega_{\operatorname{Spec} E}$ (which we shall rather denote by ω) is injective if and only if E is Hausdorff.

Proposition 5.2. – *Suppose that* $S \subset \operatorname{Spec} E$ *has the following property*:

If $\mathfrak{p}_1, \mathfrak{p}_2 \in S$ *there is* $\mathfrak{p}_3 \in S$ *such that* $\sup(\mathfrak{p}_1, \mathfrak{p}_2) \leqslant \mathfrak{p}_3$. (5.4)

Then the image of the canonical map ω_S *of* E *into* $\Gamma(S)$ *is dense.*

Proof: Let $s \in \Gamma(S)$, A be a finite subset of S, $\mathfrak{p} \in S$ such that $\mathfrak{q} \leqslant \mathfrak{p}$ for all $\mathfrak{q} \in A$; such a seminorm exists in virtue of (5.4). Since $\omega_{\mathfrak{p}}(E)$ is dense in $E_{\mathfrak{p}}$, to every $\varepsilon > 0$ there is $x \in E$ such that

$$\|s(\mathfrak{p}) - \omega_{\mathfrak{p}}(x)\|_{\mathfrak{p}} < \varepsilon .$$

Using the fact that both s and $\omega_S x$ are regular sections, and that the canonical mappings $\omega_{\mathfrak{q}}^{\mathfrak{p}} (\mathfrak{q} \in A)$ have norm $\leqslant 1$, we obtain

$$\|s(\mathfrak{q}) - \omega_{\mathfrak{q}}(x)\|_{\mathfrak{q}} < \varepsilon \quad \text{for all} \quad \mathfrak{q} \in A .$$

Q.E.D.

Proposition 5.3. – *The canonical map* ω *of* E *into* $\Gamma(\operatorname{Spec} E)$ *is a homomorphism with dense image. If* E *is Hausdorff, it is a monomorphism; if* E *is Hausdorff and complete, it is an isomorphism.*

It suffices to check that ω is a homomorphism, which is obvious on the definition of the topologies of E and $\Gamma(\operatorname{Spec} E)$.

If $S \subset S' \subset \operatorname{Spec} E$, the restriction of sections of the fibre set $\mathfrak{E}$, from S' to S, defines a continuous linear map of $\Gamma(S')$ into $\Gamma(S)$, $\varrho_S^{S'}$. Of course, we have

$$\omega_S = \varrho_S^{S'} \circ \omega_{S'}.$$

Let now F be another locally convex space, $u: E \to F$ a continuous linear map. Let S be a subset of $\operatorname{Spec} E$, s a regular section of $\mathfrak{E}$ over S. Then

$$\mathfrak{q} \rightsquigarrow u_{\mathfrak{q}}(s(u_* \mathfrak{q})) \ (\in \hat{F}_{\mathfrak{q}}) \tag{5.5}$$

is a regular section over $\bar{u}_*^1(S)$. This follows at once from Proposition 5.1. The section (5.5) will be denoted by us and called *the direct image of s under u.* We have

$$u\omega_S x = \varpi_T u(x),$$

where $T = \bar{u}_*^1 S$ and ϖ is the analog of ω for F instead of E. As $u_{\mathfrak{q}}$ is an isometry of $\hat{E}_{u_*\mathfrak{q}}$ into $\hat{F}_{\mathfrak{q}}$, we see that the direct image $s \rightsquigarrow us$ is a continuous linear mapping of $\Gamma(S)$ into $\Gamma(\bar{u}_*^1 S)$.

Examples

Example 5.1. – Let E be an arbitrary locally convex TVS; then $\hat{E}_{\mathfrak{p}}$ is a one-dimensional Hilbert space if and only if there is a linear functional x' on E such that $\mathfrak{p} = |x'|$, i.e., $\mathfrak{p}(x) = |\langle x', x\rangle|$ for all $x \in E$ (see Proposition 15.1).

Example 5.2. – Let Ω be an open subset of a Euclidean space $\mathbf{R}^n$, $E = \mathscr{C}^\infty(\Omega)$, the space of infinitely differentiable complex functions defined in Ω. Let m be any finite integer $\geqslant 0$. Consider the seminorm $\mathfrak{p}$ on E,

$$f \rightsquigarrow \left(\sum_{|\alpha| \leqslant m} \int_{\Omega'} |(\partial/\partial x)^\alpha f(x)|^2 \, dx \right)^{1/2}, \tag{5.6}$$

where Ω' is a relatively compact open subset of Ω, $\alpha = (\alpha_1, \ldots, \alpha_n)$ is an n-tuple of nonnegative integers, the α_j's, and as usual,

$$|\alpha| = \alpha_1 + \cdots + \alpha_n, \quad (\partial/\partial x)^\alpha = (\partial/\partial x_1)^{\alpha_1} \ldots (\partial/\partial x_n)^{\alpha_n}.$$

When the boundary of Ω' is smooth enough, the Banach space $\hat{E}_{\mathfrak{p}}$ can be identified with the space $H^m(\Omega')$ of distributions f in Ω' whose derivatives of order $\leqslant m$ belong to $L^2(\Omega')$ (with its natural norm, given precisely by (5.6) – $\hat{E}_{\mathfrak{p}}$ is then a Hilbert space).

Example 5.3. – Let Ω be now an open subset of the complex space $\mathbf{C}^n$ and take $E = H(\Omega)$, the space of holomorphic functions in Ω. Consider the seminorm $\mathfrak{p}$ on $E, f \rightsquigarrow \sup_{x \in K} |f(x)|$, where K is a compact subset of Ω. Under favorable circumstances, e.g., when K is the closure of its interior $\mathring{K}$ and $\mathring{K}$ is star-shaped with respect to some of its points, the Banach space $\hat{E}_\mathfrak{p}$ may be identified with the space of functions, continuous in K and holomorphic in $\mathring{K}$.

In all the examples of the kind of 5.2, 5.3, the mappings $\omega_\mathfrak{q}^\mathfrak{p}$ (with $\mathfrak{p}$, $\mathfrak{q} \in \operatorname{Spec} E$, $\mathfrak{q} \leq \mathfrak{p}$, of the type considered) are defined by restrictions of functions or distributions.

6. The irreducible subsets of the spectrum interpreted as the equivalence classes of continuous linear mappings with dense image

The reading of this section, if one excepts the last portion of it, is not strictly indispensable for the understanding of the sequel. It might however help this understanding, as it provides a simple interpretation of the spectrum of E and of its irreducible subsets, as indicated in the title.

We begin by the following result:

Proposition 6.1. – *Let E, F be locally convex spaces, $u: E \to F$ a continuous linear mapping.*

Images and preimages, under u_*, of irreducible sets are irreducible sets.

Proof: Let A (resp. B) be an irreducible subset of $\operatorname{Spec} E$ (resp., of $\operatorname{Spec} F$). Obviously, $u_*(B)$ and $\overset{-1}{u}_*(A)$ verify (2.1); they also verify (2.3) as we derive from (4.2). It remains to check (2.2).

1°) Let $\mathfrak{q}$, $\mathfrak{r} \in \operatorname{Spec} F$ be such that $u_*\mathfrak{q} \in A$ and $\mathfrak{r} \leqslant \mathfrak{q}$. Since u_* is order preserving, we have $u_*\mathfrak{r} \leqslant u_*\mathfrak{q}$, hence $u_*\mathfrak{r} \in A$, since A satisfies (2.2).

2°) Let $\mathfrak{p} \in \operatorname{Spec} E$, $\mathfrak{q} \in \operatorname{Spec} F$ be such that $\mathfrak{q} \in B$ and $\mathfrak{p} \leqslant u_*\mathfrak{q}$. Then $\mathfrak{p}$ vanishes on $\operatorname{Ker} u$ and, therefore, $u(x) \rightsquigarrow \mathfrak{p}(x)$ is a well defined seminorm on $\operatorname{Im} u$, obviously $\leqslant \mathfrak{q}$. In view of Lemma 1.1, it can be extended to a seminorm $\mathfrak{r}$ in F, $\mathfrak{r} \leqslant \mathfrak{q}$ (hence $\mathfrak{r} \in \operatorname{Spec} F$); and $\mathfrak{p} = u_*\mathfrak{r}$.

We consider a subset S of $\operatorname{Spec} E$; ω_S denotes the canonical map of E into $\Gamma(S)$, the locally convex space of the regular sections over S.

Proposition 6.2. – *The following two properties are equivalent:*

(a) *S is an irreducible set;*

(b) *ω_{S*} is a bijection of* $\operatorname{Spec} \Gamma(S)$ *onto S.*

Proof: That (b) implies (a) is evident: indeed, $\operatorname{Spec} \Gamma(S)$ is irreducible, hence also its image under ω_{S*} (Proposition 6.1). We shall prove that (a) implies (b).

Since irreducible sets have Property (5.4), ω_{S*} is injective (Propositions

4.2 and 5.2; here we view ω_{S*} as being valued in Spec$E$). Let us show that $\mathrm{Im}\,\omega_{S*} \subset S$. As $\Gamma(S)$ carries the topology of pointwise convergence in S, to an arbitrary continuous seminorm $\mathfrak{Q}$ on $\Gamma(S)$ there is a finite subset A of S and a constant $C>0$ such that, for all $s \in \Gamma(S)$,

$$\mathfrak{Q}(s) \leqslant C \sup_{\mathfrak{p} \in A} \|s(\mathfrak{p})\|_{\mathfrak{p}}.$$

We apply this to $s = \omega_S x$, $x \in E$ arbitrary; we see that

$$\omega_{S*}\mathfrak{Q}(x) \leqslant C \sup_{\mathfrak{p} \in A} \mathfrak{p}(x),$$

hence

$$\omega_{S*}\mathfrak{Q} \leqslant C \sup_{\mathfrak{p} \in A} \mathfrak{p}.$$

We must then have $\omega_{S*}\mathfrak{Q} \in S$ in virtue of (2.1), (2.2), (2.3). It is then obvious that $\mathrm{Im}\,\omega_{S*} \supset S$: the image under ω_{S*} of the seminorm on $\Gamma(S)$, $s \rightsquigarrow \|s(\mathfrak{p})\|_{\mathfrak{p}}$ $(\mathfrak{p} \in S)$, is obviously $\mathfrak{p}$. Q.E.D.

Definition 6.1. – *We denote by* **Irr**(E) *the collection of all the irreducible subsets of* Spec$E$.

We put on **Irr**(E) the order relation "inclusion of sets"; then **Irr**(E) has a maximal element, Spec$E$, and a minimal one, the set consisting of the seminorm identically equal to zero (we exclude the empty set from **Irr**(E)). As a matter of fact, **Irr**(E) is a lattice: the intersection of two irreducible sets S_1, S_2 is $\inf(S_1, S_2)$, the irreducible hull of $S_1 \cup S_2$ is $\sup(S_1, S_2)$.

Next, we consider the set of all the continuous linear mappings of E into some locally convex space *whose image is dense*. The values space is not fixed; it can vary. It will always be *Hausdorff* and *complete*. The set of such mappings will be denoted by $\mathscr{L}(E)$. Two mappings belonging to $\mathscr{L}(E)$, $u_i: E \to F_i$ $(i=1, 2)$ will be called *equivalent* if there exists an isomorphism $J: F_1 \to F_2$ such that $J \circ u_1 = u_2$, i.e., if the following diagram is commutative

$$\begin{array}{ccc} & & F_1 \\ & \nearrow^{u_1} & \\ E & & \downarrow J \\ & \searrow_{u_2} & \\ & & F_2. \end{array} \tag{6.1}$$

Definition 6.2. – *We denote by* $\mathbf{L}(E)$ *the quotient of* $\mathscr{L}(E)$ *modulo the relation "u_1 and u_2 are equivalent".*

If the mapping J, in the commutative diagram (6.1), is not an isomorphism but simply a continuous linear map (necessarily with dense image since $\operatorname{Im} u_2 \subset \operatorname{Im} J$), we say that u_1 is *stronger* than u_2 or that u_2 is *weaker* than u_1. Now, the mappings u_1 *and* u_2 *are equivalent if and only if each one is stronger than the other*. The "only if" part of the assertion is obvious. Conversely, suppose that we have a "double" commutative diagram

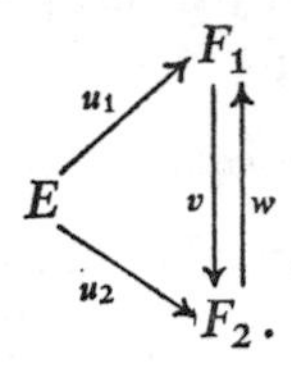

Then $u_1 = w \circ u_2 = w \circ v \circ u_1$, hence $w \circ v$ is equal to the identity map of F_1 on a dense subset of F_1, therefore everywhere. Similarly, $v \circ w$ is the identity mapping of F_2. This proves that v and w are inverse of each other and that they are isomorphisms. Q.E.D. The relation "to be stronger than" is a preorder relation on $\mathscr{L}(E)$. It induces an *order* relation on $\mathbf{L}(E)$, which we express by saying that "γ_1 is stronger than γ_2": it means that every representative of γ_1 is stronger than every representative of γ_2.

Proposition 6.3. – *The ordered set* $\mathbf{L}(E)$ *is a lattice.*

Proof: Consider two mappings belonging to $\mathscr{L}(E)$, $u_i : E \to F_i$ $(i=1, 2)$. Let $v : x \rightsquigarrow (u_1(x), u_2(x))$ map E into $F_1 \times F_2$.

1°) Let F^{12} be the closure of $\operatorname{Im} v$ in $F_1 \times F_2$; if we regard v as a continuous linear map (with dense image!) of E into F^{12}, we can check easily that its equivalence class is the *supremum* of the equivalence classes of u_1 and u_2.

2°) Let us put on $\operatorname{Im} v$ the topology defined by the seminorms

$$(y_1, y_2) \rightsquigarrow \mathfrak{q}_1(y_1) + \mathfrak{q}_2(y_2)$$

where $\mathfrak{q}_i \in \operatorname{Spec} F_i$ $(i=1, 2)$ are seminorms related by the condition

$$u_{1*}\mathfrak{q}_1 = u_{2*}\mathfrak{q}_2 .$$

Let G be the Hausdorff locally convex space associated with $\operatorname{Im} v$ so topologized, and let F_{12} be the completion of G. Let φ be the canonical map of $\operatorname{Im} v$ onto G and j the natural injection of G into F_{12}. The equivalence class of $j \circ \varphi \circ v (v : E \to \operatorname{Im} v)$ is the *infimum* of the equivalence classes of u_1 and u_2 – as can easily be checked.

We are faced, at this stage, with two lattices:

$\mathbf{Irr}(E)$, the collection of all the irreducible (Definition 2.1) subsets of $\operatorname{Spec} E$, with the inclusion of sets as order relation;

$\mathbf{L}(E)$, the set of equivalence classes (modulo isomorphisms) of continuous linear mappings, with dense image, of E into some locally convex Hausdorff complete space, the order relation being the relation "to be weaker than".

The main result of the present section is that these two lattices are canonically isomorphic. We begin by definnig the "canonical isomorphism".

Consider the first mapping:

$$\mathbf{Irr}(E) \ni S \rightsquigarrow \textit{equivalence class of } \omega_S, \tag{6.2}$$

where ω_S is the canonical map of E into $\Gamma(S)$; ω_S has a dense image in view of Proposition 5.2.

Next, consider the mapping:

$$\textit{equivalence class of } u \rightsquigarrow \operatorname{Im} u_* . \tag{6.3}$$

It is well defined: if the diagram (6.1) is commutative and if J is an isomorphism, we have $u_{2*} = u_{1*} \circ J_*$ and J_* is a bijection of $\operatorname{Spec} F_2$ onto $\operatorname{Spec} F_1$ (corollary 2 of Proposition 4.3). Hence $\operatorname{Im} u_{1*} = \operatorname{Im} u_{2*}$.

Theorem 6.1. – *The mapping* (6.2) *is a lattice isomorphism of* $\mathbf{Irr}(E)$ *onto* $\mathbf{L}(E)$. *The inverse mapping is given by* (6.3).

Proof: Let us denote momentarily by f and g respectively the mappings (6.2) and (6.3). That $g \circ f$ is the identity of $\mathbf{Irr}(E)$ is stated in Proposition 6.2. We must prove that $f \circ g$ is the identity of $\mathbf{L}(E)$.

Let $u: E \to F$ belong to $\mathscr{L}(E)$. Since $\operatorname{Spec} F$ is an irreducible set, so is $\operatorname{Im} u_*$ (Proposition 6.1); let us set $S = \operatorname{Im} u_*$. It is obvious that

$$u(x) \rightsquigarrow \omega_S(x)$$

is an isomorphism of $\operatorname{Im} u$, equipped with the topology induced by F, onto $\operatorname{Im} \omega_S$, equipped with the topology induced by $\Gamma(S)$. As both $\operatorname{Im} u$ and $\operatorname{Im} \omega_S$ are dense in the values spaces, F and $\Gamma(S)$ respectively, this isomorphism can be extended as an isomorphism of F onto $\Gamma(S)$, J; we have $\omega_S = J \circ u$, whence the equivalence of u and ω_S.

It remains to show that f and g preserve the order relations.

1°) Let $S_1, S_2 \in \mathbf{Irr}(E)$ be such that $S_1 \supset S_2$ and let $\varrho_{S_2}^{S_1}$ denote the restriction of regular sections from S_1 to S_2; we have

$$\omega_{S_2} = \varrho_{S_2}^{S_1} \circ \omega_{S_1},$$

hence ω_{S_1} is stronger than ω_{S_2}.

2°) Let $u_i: E \to F_i$ $(i = 1, 2)$ belong to $\mathscr{L}(E)$ and be such that, for some continuous linear map $v: F_1 \to F_2$, we have $u_2 = v \circ u_1$. We have then $u_{2*} = u_{1*} \circ v_*$, hence $\operatorname{Im} u_{2*} \subset \operatorname{Im} u_{1*}$.

Irreducible hulls of points of the spectrum

Let E be a vector space, $\mathfrak{p}$ a seminorm on E. We shall denote by $S_{\mathfrak{p}}$ the irreducible hull of the set $\{\mathfrak{p}\}$ (regarded as a subset of the *total* spectrum of E, that is to say, of the set of all the seminorms on E, set which we denote by SPECE). The elements of $S_{\mathfrak{p}}$ are those seminorms $\mathfrak{q}$ on E such that, for some constant $C>0$, we have $\mathfrak{q}\leqslant C\mathfrak{p}$. Now, if $\mathfrak{q}$ is such a seminorm on E, it defines a seminorm $\dot{\mathfrak{q}}$ on $E/\mathrm{Ker}\,\mathfrak{p}$ by setting

$$\dot{\mathfrak{q}}(\varphi(x))=\mathfrak{q}(x),$$

where φ is the canonical surjection of E onto $E/\mathrm{Ker}\,\mathfrak{p}$. Moreover, this seminorm $\mathfrak{q}$ is continuous on $E/\mathrm{Ker}\,\mathfrak{p}$ for the topology defined by the norm $\|\ \|_{\mathfrak{p}}$ associated with the seminorm $\mathfrak{p}$ (see p. 15). Thus, $\mathfrak{q}$ can be extended as a continuous seminorm $\hat{\mathfrak{q}}$ on the Banach space $\hat{E}_{\mathfrak{p}}$. Denoting, as before, the canonical mapping of E into $\hat{E}_{\mathfrak{p}}$ by $\omega_{\mathfrak{p}}$ we see that we have

$$(\omega_{\mathfrak{p}})_*\hat{\mathfrak{q}}=\mathfrak{q}.$$

Conversely, if r is any continuous seminorm on $\hat{E}$, $(\omega_{\mathfrak{p}})_*\mathfrak{r}$ belongs to $S_{\mathfrak{p}}$. Thus we see that $(\omega_{\mathfrak{p}})_*$ is a bijection of Spec$\hat{E}_{\mathfrak{p}}$ onto $S_{\mathfrak{p}}$. Suppose that E is now equipped with a locally convex topology for which $\mathfrak{p}$ is continuous, so that now $S_{\mathfrak{p}}\subset\mathrm{Spec}\,E$. Then, in the language of the first part of this section, we might say that $S_{\mathfrak{p}}$ is the irreducible subset of Spec$E$ corresponding to the equivalence class of $\omega_{\mathfrak{p}}$ modulo isomorphisms – in the canonical correspondence between $\mathbf{Irr}(E)$ and $\mathbf{L}(E)$.

We can define a *metric* on the set $S_{\mathfrak{p}}$, in the following manner:

$$\|\mathfrak{q}_1-\mathfrak{q}_2\|_{\mathfrak{p}}=\sup_{x\in E,\ \mathfrak{p}(x)\neq 0}\mathfrak{p}(x)^{-1}\|\mathfrak{q}_1(x)-\mathfrak{q}_2(x)\|.$$

Proposition 6.4. – *The metric space $S_{\mathfrak{p}}$ is complete.*

Proof: Let $\{\mathfrak{q}_k\}$ $(k=0,1,\ldots)$ be a Cauchy sequence in $S_{\mathfrak{p}}$; for each $x\in E$, the nonnegative numbers $\mathfrak{q}_k(x)$ converge to a number, $\mathfrak{q}(x)$; obviously, $x\rightsquigarrow\mathfrak{q}(x)$ is a seminorm on E. As the norms $\|\mathfrak{q}_k\|_{\mathfrak{p}}$ are bounded by some constant $M<+\infty$ independent of k, we have $\mathfrak{q}\leqslant M\mathfrak{p}$.

7. Spectra of inductive and projective limits of locally convex spaces

For want of a better place, we say now a few words about the spectra of a locally convex space which is either the inductive limit, or the projective one, of a family of locally convex TVS. As we shall define inductive (resp. projective) limits, topological direct sums (resp. products) are particular cases of these. We shall be dealing with the vector space E on which we are going to define a locally convex topology, and with a collection $\{E_\alpha\}$ $(\alpha \in A)$ of locally convex spaces. The set of indices A can be anything, at least in the definitions. For each index α, we are given a linear map u_α. If we are aiming at defining the inductive limit, u_α will be a mapping of E_α into E; if we are headed towards the projective limit, $u_\alpha: E \to E_\alpha$. By going to the *total* spectra, we obtain the mappings

$$u_{\alpha*}: \mathrm{SPEC}\, E \to \mathrm{SPEC}\, E_\alpha \quad \text{(for the inductive limit);}$$
$$u_{\alpha*}: \mathrm{SPEC}\, E_\alpha \to \mathrm{SPEC}\, E \quad \text{(for the projective limit).}$$

We recall that a locally convex topology on E is the datum of an irreducible subset of SPEC E. Then, the *inductive limit topology* on E defined by the spaces E_α and the mappings $u_\alpha: E_\alpha \to E (\alpha \in A)$ is defined by the following irreducible subset of SPEC E:

$$\bigcap_{\alpha \in A} \overset{-1}{u_{\alpha*}} (\mathrm{Spec}\, E_\alpha),$$

whereas the *projective limit topology* on E defined by the spaces E_α and the mappings $u_\alpha: E \to E_\alpha$, is defined by *the irreducible hull*, in SPEC E, of the set

$$\bigcup_{\alpha \in A} u_{\alpha*} (\mathrm{Spec}\, E_\alpha).$$

The reader will easily see that these definitions of projective and inductive limit topologies are identical with the usual ones.

Let now E be an arbitrary vector space, S an irreducible subset of SPEC E. One sees at once that the locally convex topology on E such that S becomes the spectrum of E is the projective limit of the topologies of the Banach spaces $\hat{E}_\mathfrak{p}$ with respect to the mappings $\omega_\mathfrak{p}$ as $\mathfrak{p}$ varies over S.

Chapter III. Epimorphisms of Fréchet Spaces

8. Equicontinuous subsets of the spectrum

Let S be a set of seminorms on E; S is said to be equicontinuous at the point x^0 of E if to every $\varepsilon > 0$ there is a neighborhood $x^0 + U$ of x^0 (U: neighborhood of O in E) such that, for all $\mathfrak{p} \in S$ and all $x \in x^0 + U$,

$$|\mathfrak{p}(x) - \mathfrak{p}(x^0)| \leqslant \varepsilon .$$

In view of the triangular inequality, this is equivalent with saying that, for all $y \in U$,

$$\mathfrak{p}(y) \leqslant \varepsilon ,$$

or, equivalently, for all $x \in \varepsilon^{-1} U$,

$$\mathfrak{p}(x) \leqslant 1 .$$

In other words, we see that S is equicontinuous if and only there is a neighborhood of O in E, V (above, $V = \varepsilon^{-1} U$), such that

$$V \subset B_{\mathfrak{p}} \quad \text{for all} \quad \mathfrak{p} \in S ,$$

i.e., if and only if the intersection of all the closed unit semiballs of the seminorms belonging to S,

$$\bigcap_{\mathfrak{p} \in S} B_{\mathfrak{p}} , \tag{8.1}$$

is a neighborhood of 0 in E. Note that (8.1) is nothing else but the polar of S, S^0. It is closed convex balanced; it is also absorbent whenever it is a neighborhood of O.

Proposition 8.1. – *The following properties of a subset S of* Spec E *are equivalent*:

(a) *S is equicontinuous*;

(b) *the polar S^0 of S is a neighborhood of O*;

(c) *there is $\mathfrak{p}_0 \in \operatorname{Spec} E$ such that $\mathfrak{p} \leqslant \mathfrak{p}_0$ for all $\mathfrak{p} \in S$.*

Proof: We have just seen that (a) and (b) are equivalent; (b) is equivalent with the fact that the gauge of S^0 is everywhere finite (as S^0 is absorbent) and is continuous: then (b) implies (c) if we take $\mathfrak{p}_0$ = gauge of

S^0. But conversely, if (c) holds (with some $\mathfrak{p}_0 \in \operatorname{Spec} E$), we have $B_{\mathfrak{p}_0} \subset B_{\mathfrak{p}}$ for all $\mathfrak{p} \in S$, hence $B_{\mathfrak{p}_0} \subset S^0$. Q.E.D.

Subsets of equicontinuous sets are equicontinuous, and so are finite unions of equicontinuous sets.

A subset S of $\operatorname{Spec} E$ is equicontinuous if and only if S is contained in the polar U^0 of a neighborhood of 0, U.

A subset V of E is a neighborhood of 0 if and only if V contains the polar of an equicontinuous subset of $\operatorname{Spec} E$.

Etc. etc.

9. Barrels. Barrelled spaces

Let E be a locally convex space.

Definition 9.1. – *A subset B of E is called a barrel if B is closed, convex, balanced, absorbent.*

The closed unit semiball of a *continuous* seminorm is a barrel. The closed unit semiball of a seminorm which is not continuous is not, in general, a barrel. But it is sometimes: in other words, the gauge of a barrel is not necessarily continuous, as the following result shows:

Proposition 9.1. – *Let S be a set of continuous seminorms on E and suppose that, for every $x \in E$,*

$$\mathfrak{p}_S(x) = \sup_{\mathfrak{p} \in S} \mathfrak{p}(x) < +\infty .$$

Then the closed unit semiball of $\mathfrak{p}_S$ is a barrel.

Proof: The fact that $\mathfrak{p}_S$ is everywhere finite implies that it is a seminorm; its closed unit semiball is equal to the intersection of the closed unit semiballs $B_{\mathfrak{p}}$ as $\mathfrak{p}$ runs over S. As every $B_{\mathfrak{p}}$ is closed in E, so is $B_{\mathfrak{p}_S}$. Q.E.D.

Thus barrels are characterized, among convex balanced absorbent subsets of E, by the fact that their gauge is the upper enveloppe of the seminorms which are continuous and at most equal to it. This can be rephrased in the following manner: a convex balanced absorbent subset of E is a barrel if and only if its gauge is lower semicontinuous.

Proposition 9.2. – *The following properties of a locally convex space E are equivalent:*

(a) *every barrel of E is a neighborhood of* 0 *in E;*

(b) *if the upper enveloppe of a family of continuous seminorms on E is everywhere finite, it is continuous;*

(c) *for a subset S of* $\operatorname{Spec} E$, *it is equivalent to be equicontinuous and to be pointwise bounded.*

We say that S is pointwise bounded if, for every $x \in E$,

$$\sup_{\mathfrak{p} \in S} \mathfrak{p}(x) < +\infty .$$

Proof: The equivalence of (a) and (b) follows at once from Proposition 9.1. On the other hand, it is equivalent to say that the upper enveloppe of $S \subset \operatorname{Spec} E$ is everywhere finite or that S is pointwise bounded. In view of Proposition 8.1, it is also equivalent to say that S is equicontinuous or that the upper enveloppe of S is a continuous seminorm on E. Q.E.D.

Definition 9.2. – *A locally convex space E endowed with the equivalent properties* (a), (b), (c) *in Proposition* 9.2 *is said to be barrelled.*

Quotient spaces and completions of barrelled spaces are barrelled. But a vector subspace of a barrelled space is not necessarily barrelled (even if the subspace is closed!). Products, inductive limits of barrelled spaces are barrelled. The topological direct sum of a collection of locally convex spaces E_α is barrelled if and only if every E_α is barrelled.

We recall that a topological space is said to be a *Baire space* if countable intersections of everywhere dense open sets are everywhere dense sets (equivalently, if countable unions of closed sets without interior points have no interior points).

Proposition 9.3. – *A locally convex space E which is a Baire space is barrelled.*

Proof: Let B be a barrel in E; as B is balanced and absorbent, $E = \bigcup_{n=1}^{\infty} nB$, hence some nB, and therefore also B, must have an interior point. But the interior of a convex balanced set is also convex and balanced, hence the origin must be an interior point of B. Q.E.D.

Corollary. – *Fréchet and Banach spaces are barrelled.*

Fréchet spaces are locally convex metrizable complete TVS. There are normed spaces which are not barrelled, and there are barrelled spaces which are not metrizable (e.g., arbitrary products of barrelled spaces).

10. The epimorphism theorem

We consider two locally convex spaces E, F and a continuous linear map $u: E \to F$. We are concerned, now, with the important problem of finding conditions which insure that u is surjective – and which can be used in the applications. It needs not be emphasized that this is an important problem: it is the problem of showing that a functional equation

$u(x)=y$ has at least one solution x for every y in the "data space" F. Also, it needs hardly saying that we shall not solve this problem in full generality! In the present section, we shall find conditions, both necessary and sufficient, in order that u be surjective, in the particular (but important) case where E and F are Fréchet spaces. Later on, we shall enlarge the stock of spaces for which such conditions can be found. Our treatment is directed primarily towards obtaining conditions which are directly applicable to the theory of linear partial differential equations. As a matter of fact, the whole present work stems from the observation that linear PDE theory uses conditions of surjectivity (in the form of "*a priori* estimates", local solvability or the so-called **P**-convexity) which are not immediate translations of the "classical" surjectivity conditions (due to S. BANACH) but which, however, have natural counterparts in the abstract situation. And, as it often happens in mathematics, these new abstract criteria of surjectivity are quite simple to state – of course, much simpler to use – but also much simpler to prove than the old criteria.

Suppose momentarily that E and F are complete locally convex spaces. Then we may interpret them as the spaces of regular sections over their respective spectra (Proposition 5.3); we may also interpret the map u as the direct image of regular sections (see p. 18). Thus, to say that u is surjective is equivalent with saying that, to every regular section t over $\operatorname{Spec} F$, there is a regular section s over $\operatorname{Spec} E$ such that $us=t$.

We go back, now, to the spaces $\hat{E}_{\mathfrak{p}}$ and $\hat{F}_{\mathfrak{q}}$, and to the mappings $u_{\mathfrak{q}}$ and $u_{\mathfrak{q}}^{\mathfrak{p}}$ introduced in Section 5. An immediate consequence of the surjectivity of u is that, given arbitrarily $\mathfrak{q}\in\operatorname{Spec} F$ and $\hat{y}\in\hat{F}_{\mathfrak{q}}$, we can find $\hat{x}\in\hat{E}_{u*\mathfrak{q}}$ such that $u_{\mathfrak{q}}(\hat{x})=\hat{y}$ (for the proof of this assertion, see Proposition 10.1 below). One could then ask, perhaps too optimistically, if the latter property does not imply the a-priori stronger one, that u itself is surjective. The following result shows that this is too much asking, at least in general:

Proposition 10.1. – *The following properties are equivalent*:

(a) $\operatorname{Im} u$ *is dense in* F;

(b) *for every* $\mathfrak{q}\in\operatorname{Spec} F$, $u_{\mathfrak{q}}:\hat{E}_{u_*\mathfrak{q}}\to\hat{F}_{\mathfrak{q}}$ *is surjective.*

Proof: We use the notation in Diagram (5.3), recalling that $u_{\mathfrak{q}}^{\mathfrak{p}}=u_{\mathfrak{q}}$ when $\mathfrak{p}=u_*\mathfrak{q}$. Property (a) is equivalent with saying that, for every $\mathfrak{q}\in\operatorname{Spec} F$, $\operatorname{Im} u$ is dense in $F_{\mathfrak{q}}$, i.e., $u_{\mathfrak{q}}(\omega_{u_*\mathfrak{q}}(E))$, which is always contained in $\varpi_{\mathfrak{q}}(F)$, is dense in $\varpi_{\mathfrak{q}}(F)$. As $\varpi_{\mathfrak{q}}(F)$ is dense in $\hat{F}_{\mathfrak{q}}$, we see that (a) is equivalent with saying that $u_{\mathfrak{q}}(\omega_{u_*\mathfrak{q}}(E))$ is dense in $\hat{F}_{\mathfrak{q}}$. As $\omega_{u_*\mathfrak{q}}(E)$ is dense in $E_{u_*\mathfrak{q}}$, this is equivalent with saying that $\operatorname{Im} u_{\mathfrak{q}}$ is dense. But $u_{\mathfrak{q}}$ is an isometry: $\operatorname{Im} u_{\mathfrak{q}}$ is dense if and only if $u_{\mathfrak{q}}$ is surjective. Q.E.D.

Thus, surjectivity of every mapping $u_{\mathfrak{q}}$ is indeed a weaker property than the surjectivity of u. We shall see that, in the particular instance where E and F are Fréchet space, an intermediary property turns out to be equivalent with the surjectivity of u; this intermediary property is now defined:

Definition 10.1. – *We say that the mapping u is presurjective if u verifies the following condition*:

(PS) *For every equicontinuous subset A of* Spec E *and every $y \in F$, there is a regular section s over A such that*

$$us = \varpi_B(y), \quad B = \overset{-1}{u_*}(A).$$

We recall that $\varpi_B(y)$ is the regular section over B defined by y (ϖ is the analog of ω when F is substituted for E; see p. 17).

Proposition 10.2. – *A surjective continuous linear map is presurjective.*

Proof: If u is surjective, to every $y \in F$, we can find $x \in E$ such that $u(x) = y$. Then, whatever be the subset A of Spec E, we have

$$u\omega_A(x) = \varpi_B(y), \quad B = \overset{-1}{u_*}(A).$$

Proposition 10.3. – *The following properties are equivalent*:

(a) *u is presurjective*;

(b) *for every neighborhood U of O in E, the closure $\overline{u(U)}$ of $u(U)$ contains a barrel in F*;

(c) *the preimage, under u_*, of any equicontinuous subset of* Spec E *is pointwise bounded.*

We recall that a subset B of Spec F is said to be pointwise bounded if, for all $y \in F$,

$$\sup_{\mathfrak{q} \in B} \mathfrak{q}(y) < +\infty.$$

Proof of Proposition 10.3: (a) implies (c). Indeed, let A be an equicontinuous subset of Spec E, B its preimage under u_*. In view of Proposition 8.1, we may assume that A consists exactly of all the seminorms $\mathfrak{p}$ on E which are at most equal to a given seminorm $\mathfrak{p}_0 \in$ Spec E. Let $y \in F$ be arbitrary, and $s \in \Gamma(A)$, the space of regular sections over A, be such that $us = \varpi_B(y)$. For all $\mathfrak{q} \in B$, i.e., all $\mathfrak{q} \in$ Spec F such that $u_*\mathfrak{q} \leqslant \mathfrak{p}_0$, we have:

$$\|s(u_*\mathfrak{q})\|_{u_*\mathfrak{q}} \leqslant \|s(\mathfrak{p}_0)\|_{\mathfrak{p}_0};$$

but

$$us(\mathfrak{q}) = u_{\mathfrak{q}}[s(u_*\mathfrak{q})] = \varpi_{\mathfrak{q}}(y)$$

(for the meaning of $\varpi_{\mathfrak{q}}$, see Diagram (5.3)). We have also:

$$\|us(\mathfrak{q})\|_{\mathfrak{q}} = \|s(u_*\mathfrak{q})\|_{u_*\mathfrak{q}} = \|\varpi_{\mathfrak{q}}(y)\|_{\mathfrak{q}}$$

since $u_{\mathfrak{q}}$ is an isometry; also $\|\varpi_{\mathfrak{q}}(y)\|_{\mathfrak{q}}=\mathfrak{q}(y)$. Finally, we see that, for every $y \in F$,

$$\mathfrak{q}(y) \leqslant \|s(\mathfrak{p}_0)\|_{\mathfrak{p}_0}.$$

This implies at once (c).

(c) implies (b). It suffices to prove that $\overline{u(U)}$ is a barrel whenever U is the closed unit semiball of a continuous seminorm $\mathfrak{p}_0$ on E. Let B be the subset of Spec F consisting of the seminorms $\mathfrak{q}$ such that $u_*\mathfrak{q} \leqslant \mathfrak{p}_0$. In view of Proposition 8.1 and of Property (c) (which is supposed to hold), we have $\mathfrak{q}_0(y)=\sup_{\mathfrak{q}\in B}\mathfrak{q}(y)<+\infty$ for all $y\in F$. I contend that $\overline{u(U)}$ is equal to the closed unit semiball $B_{\mathfrak{q}_0}$ of the seminorm $\mathfrak{q}_0$; by Proposition 9.1, this implies (b). Let $x\in E$ be such that $\mathfrak{p}_0(x)\leqslant 1$; then $\mathfrak{q}(u(x))\leqslant 1$ for all $\mathfrak{q}\in B$, hence $\mathfrak{q}_0(u(x))\leqslant 1$, hence $u(U)$, and therefore also $\overline{u(U)}$, is contained in $B_{\mathfrak{q}_0}$. Let $y\in F$ belong to the complement of $\overline{u(U)}$. We apply Lemma 3.1: there is a seminorm $\mathfrak{q}\in \operatorname{Spec} F$, $\mathfrak{q}\leqslant 1$ on $u(U)$ and $\mathfrak{q}(y)>1$; the first property implies that $u_*\mathfrak{q}\leqslant\mathfrak{p}_0$, i.e., that $\mathfrak{q}\in B$, the second one, combined with the fact that $\mathfrak{q}\in B$, implies that $\mathfrak{q}_0(y)>1$, i.e., $y\notin B_{\mathfrak{q}_0}$. Thus $B_{\mathfrak{q}_0}=\overline{u(U)}$.

(b) implies (a). As before, let $\mathfrak{p}_0\in\operatorname{Spec} E$ be arbitrary, $A\subset\operatorname{Spec} E$ consist of the seminorms $\mathfrak{p}\leqslant\mathfrak{p}_0$. Let $\mathfrak{q}_0$ be the gauge of the barrel $\overline{u(B_{\mathfrak{p}_0})}$. Now, a continuous seminorm $\mathfrak{q}$ on F verifies $u_*\mathfrak{q}\leqslant\mathfrak{p}_0$ if and only if $\mathfrak{q}\leqslant 1$ on $u(B_{\mathfrak{p}_0})$, i.e., $\mathfrak{q}\leqslant 1$ on $\overline{u(B_{\mathfrak{p}_0})}$, i.e., $\mathfrak{q}\leqslant\mathfrak{q}_0$. As we have obviously $\mathfrak{q}_0\circ u\leqslant\mathfrak{p}_0$, we may consider the mapping

$$u_{\mathfrak{q}_0}^{\mathfrak{p}_0}\colon \hat{E}_{\mathfrak{p}_0}\to\hat{F}_{\mathfrak{q}_0} \qquad \text{(see p. 16)}.$$

This is a continuous linear map of a Banach space into another one which, by our definition of $\mathfrak{q}_0$, has the following property: it maps the closed unit ball onto a dense subset of the closed unit ball. We apply Lemma A.1 (Appendix A) and conclude that this mapping is, in fact, an epimorphism. This will easily imply (a). Let $y\in F$ be arbitrary. There is $\hat{x}\in\hat{E}_{\mathfrak{p}_0}$ such that $u_{\mathfrak{q}_0}^{\mathfrak{p}_0}(\hat{x})=\varpi_{\mathfrak{q}_0}(y)$ (cf. Diagram (5.3)). Let s be the section on A defined by

$$s(\mathfrak{p})=\omega_{\mathfrak{p}}^{\mathfrak{p}_0}(\hat{x}).$$

Setting $B=\overset{-1}{u_*}(A)$, we derive from Proposition 5.1, for all $\mathfrak{q}\in B$,

$$u_{\mathfrak{q}}[s(u_*\mathfrak{q})]=u_{\mathfrak{q}}[\omega_{u_*\mathfrak{q}}^{\mathfrak{p}_0}(\hat{x})]=\varpi_{\mathfrak{q}}^{\mathfrak{q}_0}(u_{\mathfrak{q}_0}^{\mathfrak{p}_0}(\hat{x}))=\varpi_{\mathfrak{q}}(y),$$

which means precisely that $us=\varpi_B(y)$. In order to see that we have indeed obtained (a), it suffices to observe that every equicontinuous subset

S of $\operatorname{Spec} E$ is contained in some set A of the type considered above (by Proposition 8.1) and that, whatever be $y \in F$, $\varpi_T(y)$, where $T = \bar{u}_*^{-1}(S)$, is equal to the restriction to T of $\varpi_B(y)$, where $B = \bar{u}_*^{-1}(A)$ as before.

Corollary 1. – *The image of a presurjective continuous linear map is dense.*

Proof: Let $u: E \to F$ be presurjective and U be some neighborhood of O in E; then $\overline{u(U)}$ is absorbent, hence the vector subspace of F spanned by $u(U)$, which is precisely $\operatorname{Im} u$, must be dense.

Proposition 10.4. – *The following properties are equivalent:*

(b′) *for every neighborhood U of O in E, the closure $\overline{u(U)}$ of $u(U)$ is a neighborhood of O in F;*

(c′) *the preimage, under u_*, of any equicontinuous subset of* $\operatorname{Spec} E$ *is an equicontinuous subset of* $\operatorname{Spec} F$.

They imply the following property:

(a′) *For every equicontinuous subset A of* $\operatorname{Spec} E$ *and every regular section t over $B = \bar{u}_*^{-1}(A)$, there is a regular section s over A such that*

$$us = t.$$

Furthermore, if F is barrelled, (a′), (b′), (c′) *are equivalent with the fact that u is presurjective.*

Property (a′) states that the direct image under u is a surjection of $\Gamma(A)$, the space of regular sections over A, onto $\Gamma(B)$.

Proof of Proposition 10.4: The implications (c′)⇒(b′)⇒(a′) are proved exactly as the implications (c)⇒(b)⇒(a) in Proposition 10.3, with the only difference that the seminorm on F denoted by $\mathfrak{q}_0$ in the proof of Proposition 10.3 is now continuous in F. When F is barrelled, Conditions (b), Proposition 10.3, and (b′), Proposition 10.4, are identical. On the other hand, (a′) implies trivially that u is presurjective. It will therefore suffice to prove that (b′)⇒(c′).

In view of Proposition 8.1, it suffices to prove that, if (b′) holds, for every seminorm $\mathfrak{p}_0 \in \operatorname{Spec} E$ there is a seminorm $\mathfrak{q}_0 \in \operatorname{Spec} F$ such that, for all $\mathfrak{q} \in \operatorname{Spec} F$ verifying $u_* \mathfrak{q} \leqslant \mathfrak{p}_0$, we have $\mathfrak{q} \leqslant \mathfrak{q}_0$. It suffices indeed to take, as seminorm $\mathfrak{q}_0$, the gauge of $\overline{u(B_{\mathfrak{p}_0})}$: in virtue of (b′), $\mathfrak{q}_0$ is a continuous seminorm on F. Q.E.D.

Corollary 1. – *Suppose that u is a homomorphism with dense image. Then the preimage, under u_*, of any equicontinuous subset of* $\operatorname{Spec} E$ *is an equicontinuous subset of* $\operatorname{Spec} F$.

Proof: If u is a homomorphism, for every neighborhood of O, U, in E, $u(U)$ is a neighborhood of O in $\operatorname{Im} u$. Then, there exists a neighborhood of O, V, in F, such that $V \cap \operatorname{Im} u \subset u(U)$. We may take V open; then, if

$\operatorname{Im} u$ is dense, we see that $\bar{V} = \bar{V} \cap \overline{\operatorname{Im} u} \subset \overline{V \cap \operatorname{Im} u} \subset \overline{u(U)}$. This shows that (b′), in Proposition 10.4, holds. Q.E.D.

At this stage, it is helpful to summarize the main implications obtained so far:

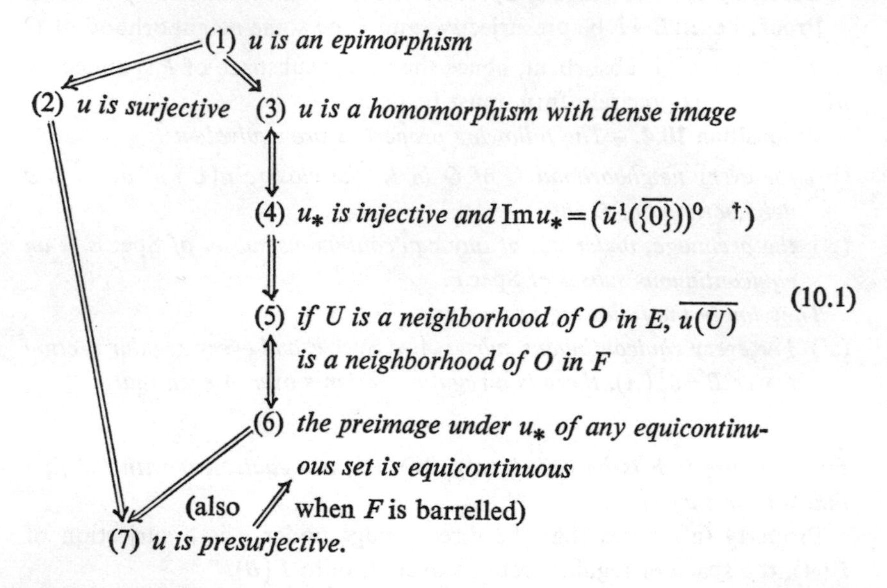

If F is a Fréchet space, F is barrelled (corollary of Proposition 9.3), whence the equivalence of (5), (6), (7). If, in addition, E also is a Fréchet space, we may apply Lemma A.1: we see that (5) implies (1). We may then state:

Theorem 10.1. – *Let E and F be two Fréchet spaces, $u: E \to F$ a continuous linear map.*

Then, Properties (1) *to* (7) *in Diagram* (10.1) *are equivalent.*

11. Criteria of presurjectivity

In this section, we consider two locally convex TVS, E and F, and a continuous linear map $u: E \to F$. We shall state and prove two conditions, sufficient in order that u be presurjective (Definition 10.1). Later on, these conditions will be applied to some situations in Linear PDE theory.

Here is our first criterion:

---

†) The equivalence of (3) and (4) is stated in Corollary 2 of Proposition 4.7; when F is Hausdorff, (4) reads:

$$u_* \text{ is injective and } \operatorname{Im} u_* = (\operatorname{Ker} u)^0. \tag{4'}$$

Proposition 11.1. – *Suppose that there is a basis of continuous seminorms on E (resp. on F), $\mathfrak{B}$ (resp. $\mathfrak{B}'$) such that the following is true:*

To every $\mathfrak{p}\in\mathfrak{B}$ there is $\mathfrak{q}\in\mathfrak{B}'$ and a constant $C>0$ such that, for all $\mathfrak{r}\in\operatorname{Spec}F$, the fact that $u_*\mathfrak{r}\in S_\mathfrak{p}$ implies that $\mathfrak{r}\in S_\mathfrak{q}$ and that (11.1)

$$\|\mathfrak{r}\|_\mathfrak{q} \leqslant C\,\|u_*\mathfrak{r}\|_\mathfrak{p}. \tag{11.2}$$

Then u is presurjective. Conversely, if F is barrelled and if u is presurjective, then (11.1) *holds for any pair $\mathfrak{B}$, $\mathfrak{B}'$ of bases of continuous seminorms on E and on F respectively.*

For the meaning of $S_\mathfrak{p}$, $S_\mathfrak{q}$, $\|\ \|_\mathfrak{p}$, $\|\ \|_\mathfrak{q}$, see p. 23.

Proof: Assume that (11.1) holds for some pair $\mathfrak{B}$, $\mathfrak{B}'$. We show then that preimages of equicontinuous subsets of $\operatorname{Spec}E$, under u_*, are equicontinuous subsets of $\operatorname{Spec}F$:

Let $\mathfrak{p}_1\in\operatorname{Spec}E$ be arbitrary; let us select $\mathfrak{p}\in\mathfrak{B}$ and a constant $M>0$ such that $\mathfrak{p}_1\leqslant M\mathfrak{p}$; let us associate with $\mathfrak{p}$ the seminorm $\mathfrak{q}\in\mathfrak{B}'$ as in (11.1), and let $\mathfrak{r}\in\operatorname{Spec}F$ be such that $u_*\mathfrak{r}\leqslant\mathfrak{p}_1$. This implies that $u_*\mathfrak{r}\in S_\mathfrak{p}$ hence $\mathfrak{r}\in S_\mathfrak{q}$ and we have

$$\|\mathfrak{r}\|_\mathfrak{q}\leqslant C\,\|u_*\mathfrak{r}\|_\mathfrak{p}\leqslant CM\,,\quad \text{i.e.,}\quad \mathfrak{r}\leqslant CM\mathfrak{q}\,.$$

In other words, $u_*\mathfrak{r}\leqslant\mathfrak{p}_1$ implies $\mathfrak{r}\leqslant\mathfrak{q}_1$ if we take $\mathfrak{q}_1=CM\mathfrak{q}$.

Assume now that preimages under u_* of equicontinuous sets are equicontinuous:

Let $\mathfrak{B}$ (resp. $\mathfrak{B}'$) be any basis of continuous seminorms on E (resp. on F), and let $\mathfrak{p}\in\mathfrak{B}$. There is a seminorm $\mathfrak{q}\in\mathfrak{B}'$ and a constant $C>0$ such that, for all $\mathfrak{r}\in\operatorname{Spec}F$,

$$u_*\mathfrak{r}\leqslant\mathfrak{p}\quad\text{implies}\quad \mathfrak{r}\leqslant C\mathfrak{q}\,,\quad\text{i.e.,}\quad \mathfrak{r}\in S_\mathfrak{q}\quad\text{and}\quad \|\mathfrak{r}\|_\mathfrak{q}\leqslant C\,.$$

Suppose, more generally, that $u_*\mathfrak{r}\in S_\mathfrak{p}$; then $\|u_*\mathfrak{r}\|_\mathfrak{p}^{-1}\,u_*\mathfrak{r}\leqslant\mathfrak{p}$, therefore $\mathfrak{r}\in S_\mathfrak{q}$ and $\big\|\,\|u_*\mathfrak{r}\|_\mathfrak{p}^{-1}\,\mathfrak{r}\big\|_\mathfrak{q}\leqslant C$, the latter implying at once (11.2). Q.E.D.

Our second criterion calls for two definitions, directly inspired by the Linear PDE theory.

Definition 11.1. – *We say that the pair (E, F) is u-convex if to every seminorm $\mathfrak{p}\in\operatorname{Spec}E$ there is a seminorm $\mathfrak{q}\in\operatorname{Spec}F$ such that, for all $\mathfrak{r}\in\operatorname{Spec}F$,*

$$u_*\mathfrak{r}\leqslant\mathfrak{p}\quad\textit{implies}\quad \operatorname{Ker}\mathfrak{q}\subset\operatorname{Ker}\mathfrak{r}\,.$$

Definition 11.2. – *We say that the mapping u is semiglobally surjective if, for every seminorm $\mathfrak{q}\in\operatorname{Spec}F$ and every $y\in F$, there is $x\in E$ such that*

$$u(x)-y\in\operatorname{Ker}\mathfrak{q}\,.$$

Here is our second criterion:

Proposition 11.2. – *If the pair (E, F) is u-convex and if u is semiglobally surjective, then u is presurjective.*

Proof: Let B be the subset of $\operatorname{Spec} F$ consisting of the seminorms $\mathfrak{r}$ such that $u_*\mathfrak{r} \leqslant \mathfrak{p}$. Since the pair (E, F) is u-convex, we have $\mathfrak{r}=0$ on $\operatorname{Ker}\mathfrak{q}$ for all $\mathfrak{r} \in B$. Let $y \in F$ be arbitrary, and let $x \in E$ be such that $u(x)-y \in \operatorname{Ker}\mathfrak{q}$. A fortiori, $u(x)-y \in \operatorname{Ker}\mathfrak{r}$ for all $\mathfrak{r} \in B$, hence (with the notation of Diagram (5.3)) $\varpi_{\mathfrak{r}}(u(x))=\varpi_{\mathfrak{r}}(y)$. If $A=\{\mathfrak{p}_1 \in \operatorname{Spec} E;\ \mathfrak{p}_1 \leqslant \mathfrak{p}\}$, we see that $u\omega_A(x)=\varpi_B(y)$.

Corollary 1. – *Let E and F be Fréchet spaces. Then $u: E \to F$ is surjective if and only if the pair (E, F) is u-convex and u is semiglobally surjective.*

Proof: If u is surjective, it is a fortiori semiglobally surjective. When F is barrelled, the presurjectivity of u implies that to every $\mathfrak{p} \in \operatorname{Spec} E$ there is $\mathfrak{q} \in \operatorname{Spec} F$ such that, for all $\mathfrak{r} \in \operatorname{Spec} F$, $u_*\mathfrak{r} \leqslant \mathfrak{p}$ implies $\mathfrak{r} \leqslant \mathfrak{q}$; the latter implies $\operatorname{Ker}\mathfrak{q} \subset \operatorname{Ker}\mathfrak{r}$, hence the pair (E, F) is u-convex. When E and F are Fréchet spaces, presurjectivity is equivalent with surjectivity (Theorem 10.1). Q.E.D.

Proposition 11.2 calls for some comments. First of all, as we have just said, the *u-convexity* of the pair (E, F) follows at once from the (stronger) property that preimages under u_* of equicontinuous subsets of $\operatorname{Spec} E$ are equicontinuous subsets of $\operatorname{Spec} F$. Therefore, it also follows from Property (11.1). The usefulness of the *semiglobal surjectivity* of u lies in the fact that, in many instances, this property is easier to establish than (11.1) (in other instances, it is the other way around). When linear PDE are concerned, semiglobal surjectivity is related, often, with the existence and certain properties of the "fundamental kernels"; this will be better understood when we discuss the examples, in Part II. Let us say, right now, that the criterion provided by Proposition 11.2, although useful in certain cases, is totally useless in others, e.g., when there is, on F, one continuous norm (in which case F possesses a basis of continuous seminorms consisting of norms). Then, if $\mathfrak{q}$ is a continuous norm on F, $\operatorname{Ker}\mathfrak{q} \subset \operatorname{Ker}\mathfrak{r}$ for all $\mathfrak{r} \in \operatorname{Spec} F$, hence, in this case, it is trivially true that the pair (E, F) is u-convex. On the other hand, in the same case, semiglobal surjectivity coincides with surjectivity!

12. The closed graph theorem

From Theorem 10.1, one derives in classical fashion the *closed graph theorem* for Fréchet spaces:

Theorem 12.1. – *Let E, F be two Frechet spaces, $u: E \to F$ a linear map. If the graph of u,*

$$Gr\,u = \{(x, y) \in E \times F;\ y = u(x)\},$$

is closed in $E \times F$, u is continuous.

Proof: Let $\mathcal{O}$ be an open subset of F; the set $(E \times \mathcal{O}) \cap \mathrm{Gr}\,u$ is open in $\mathrm{Gr}\,u$. On the other hand, the first coordinate projection $(x, y) \rightsquigarrow x$ is a continuous linear map of the Fréchet space $\mathrm{Gr}\,u$ (we assume that $\mathrm{Gr}\,u$ is closed in $E \times F$) onto E, hence is an open mapping, by Theorem 10.1: the image of $(E \times \mathcal{O}) \cap \mathrm{Gr}\,u$ under this map is open; but it is equal to $\bar{u}^1(\mathcal{O})$. Q.E.D.

We shall state and prove an easy extension of Theorem 12.1. We shall say that a locally convex Hausdorff space E is an *inductive limit of Fréchet spaces* if there is a collection of Fréchet spaces $\{E_\alpha\}$ ($\alpha \in A$) and, for each index α, a continuous linear map $\varphi_\alpha: E_\alpha \to E$, such that the following is true:

Given any locally convex space F and any linear map $u: E \to F$, u is continuous if and only if, for every $\alpha \in A$, $u \circ \varphi_\alpha: E_\alpha \to F$ is continuous. (12.1)

Corollary 1. – *Let E be an inductive limit of Fréchet spaces, F a Fréchet space, $u: E \to F$ a linear map. If the graph of u is closed, u is continuous.*

It suffices to observe that, for every Fréchet space E_0 such that there is a mapping $\varphi_0: E_0 \to E$, which is linear and continuous, the graph of $u \circ \varphi_0$ is closed in $E_0 \times F$, therefore $u \circ \varphi_0$ is continuous.

Examples of inductive limits of Fréchet spaces:

Example 12.1. – Let $\{E_k\}$ ($k = 0, 1, \ldots$) be a sequence of vector subspaces of a vector space E; we suppose that $E_k \subset E_{k+1}$ for all k, and that each E_k is equipped with a Fréchet space structure such that the natural injection of E_k into E_{k+1} is a monomorphism. We define the topology of E by requiring that a seminorm on E be continuous if and only if its restriction to each E_k is continuous. Then E is called a *space LF*. It is an inductive limit of Fréchet spaces, as one sees immediately. It is a Fréchet space if and only if $E_k = E_{k+p}$ for some k and for all $p = 1, 2, \ldots$.

Example 12.2. – We recall that a subset B of E is *bounded* if every continuous seminorm $\mathfrak{p}$ on E is bounded on B. The strong dual topology on the dual E' of E is the locally convex Hausdorff topology defined by the seminorms

$$x' \rightsquigarrow \mathfrak{p}_B(x') = \sup_{x \in B} |\langle x', x \rangle|,$$

as B varies over the family of all the bounded subsets of E. When E' carries this topology, we shall refer to it as the *strong dual* of E and denote it by E'_b. The space E is called *reflexive* if the mapping $x \rightsquigarrow (x' \rightsquigarrow \langle x', x \rangle)$ is an isomorphism, for the TVS structure, of E onto the strong dual $(E'_b)'_b$ of its own strong dual.

Let now $\mathfrak{p} \in \operatorname{Spec} E$ be arbitrary. The canonical map $\omega_\mathfrak{p}: E \to \hat{E}_\mathfrak{p}$ has a transpose ${}^t\omega_\mathfrak{p}: \hat{E}'_\mathfrak{p} \to E'$ which is injective, as $\operatorname{Im} \omega_\mathfrak{p}$ is dense. It is checked at once that this transpose is continuous when E' carries the strong dual topology, and $\hat{E}'_\mathfrak{p}$ carries the dual Banach space structure. Let $U_\mathfrak{p}$ denote the closed unit semiball of $\mathfrak{p}$, $U^0_\mathfrak{p}$ its polar in E', that is, $U^0_\mathfrak{p} = \{x' \in E'; |\langle x', x \rangle| \leq 1 \text{ for all } x \in V_\mathfrak{p}\}$; as a Banach space, $\hat{E}'_\mathfrak{p}$ can be identified, via ${}^t\omega_\mathfrak{p}$, with the subspace of E' spanned by $U^0_\mathfrak{p}$ and equipped with the norm

$$\mathfrak{p}'(x') = \inf_{x' \in \varrho U^0_\mathfrak{p}} \varrho .$$

Let now v be a linear map of E' into a locally convex space F; suppose that, for every $\mathfrak{p} \in \operatorname{Spec} E$, $v \circ ({}^t\omega_\mathfrak{p}): \hat{E}'_\mathfrak{p} \to F$ is continuous. Let V be a closed convex balanced subset of F, $U' = \overset{-1}{v}(V) \subset E'$ and $U'_\mathfrak{p} = {}^t\overset{-1}{\omega}_\mathfrak{p}(U')$. For every $\mathfrak{p} \in \operatorname{Spec} E$, $U'_\mathfrak{p}$ is a closed convex balanced neighborhood of 0 in $\hat{E}'_\mathfrak{p}$.

Let us suppose, for the remainder of the discussion, that E is a *reflexive Fréchet space*. From the fact that E is a Fréchet space and from a well known criterion of Banach (see BOURBAKI [*1*], Ch. IV, §2, n° 6, Th. 5), it follows easily that U' is closed. Because of this, U' is equal to its bipolar (for the duality with E), that is to say, to the polar of its own polar, U'^0. If we show that the latter is a bounded subset of E, we will have proved that U' is a neighborhood of 0 in E'_b, hence that v is continuous. This will mean that E'_b is the inductive limit of the spaces $\hat{E}'_\mathfrak{p}$, $\mathfrak{p} \in \operatorname{Spec} E$, hence that E'_b is an inductive limit of Fréchet spaces.

Now, to every $\mathfrak{p} \in \operatorname{Spec} E$ there is a number $\lambda_\mathfrak{p} > 0$ such that $\lambda_\mathfrak{p} U'_\mathfrak{p}$ contains the closed unit ball in the Banach space $\hat{E}'_\mathfrak{p}$, in other words, $\lambda_\mathfrak{p} U' \supset U^0_\mathfrak{p}$. If we go to the polars, we see that $U'^0 \subset \lambda_\mathfrak{p} U^0_\mathfrak{p}$, which means that $\mathfrak{p}(x) \leq \lambda'_\mathfrak{p}$ for all $x \in U'^0$. Q.E.D.

Chapter IV. Existence and Approximation of Solutions to a Functional Equation

13. The canonical extension of an essentially univalent linear mapping

In this section, we deal with four locally convex Hausdorff TVS, E_0, F_0, E, F, and with three continuous linear mappings u_0, ϱ, σ:

$$\begin{array}{ccc} E_0 & \xrightarrow{\varrho} & E \\ {\scriptstyle u_0}\downarrow & & \\ F_0 & \xrightarrow[\sigma]{} & F. \end{array} \tag{13.1}$$

Our purpose is to show that, under a certain condition, there exists a linear subspace $\tilde{E}_0$ of E, containing (as a dense subspace) $\varrho(E_0)$, and a linear map $\tilde{u}_0 : \tilde{E}_0 \to F$, such as to have

$$\tilde{u}_0 \circ \varrho = \sigma \circ u_0 . \tag{13.2}$$

We shall furthermore require that *the graph of $\tilde{u}_0$ in $E \times F$ be closed.*

First of all, we define the condition under which this is possible:

Definition 13.1. – *We say that u_0 is essentially univalent with respect to the pair of mappings (ϱ, σ), if the following holds:*

For every filter $\mathscr{F}$ on E_0, if $\varrho(\mathscr{F})$ converges to O in E and if $\sigma(u_0(\mathscr{F}))$ converges in F, we have (13.3)

$$\lim \sigma(u_0(\mathscr{F})) = 0 .$$

Suppose that (13.3) is verified. Take for $\mathscr{F}$ the filter spanned by the set consisting of a single element x_0 which satisfies $\varrho(x_0) = 0$. Then we must have $\sigma(u_0(x_0)) = 0$. Notice that this condition is certainly implied by (13.2). Of course, (13.3) contains more than this condition. We are going to see what (13.3) means – in more explicit terms. But before let us give a sufficient condition in order that (13.3) be verified:

Proposition 13.1. – *Suppose that the following diagram is commutative:*

$$\begin{array}{ccccc} E_0 & \xrightarrow{\varrho} & E & & \\ {\scriptstyle u_0}\downarrow & & & \searrow^{u} & \\ F_0 & \xrightarrow[\sigma]{} & F & \xrightarrow[j]{} & \Omega , \end{array}$$

where Ω is a locally convex Hausdorff TVS, u and j are continuous linear mappings.

If j is injective, u_0 is essentially univalent with respect to (ϱ, σ).

Proof: Let $\mathscr{F}$ be a filter on E_0. Assume that $\varrho(\mathscr{F})$ converges to 0 in E and that $\sigma(u_0(\mathscr{F}))$ converges, in F, to some $y \in F$; then $u(\varrho(\mathscr{F}))$ and $j(\sigma(u_0(\mathscr{F}))$ converge to the same element in Ω. This element is 0 on one hand, $j(y)$ on the other; therefore $j(y)=0$, hence $y=0$ as j is injective.

Q.E.D.

From now on, we omit the mention "with respect to the pair (ϱ, σ)". No ambiguity will follow from this omission.

Let us denote by G_0 the linear subspace of $E \times F$ consisting of the pairs $(\varrho(x_0), \sigma(u_0(x_0)))$, $x_0 \in E_0$. Let G be the closure of G_0 (in the product TVS $E \times F$): G is a Hausdorff locally convex space. Now, *the fact that u_0 is essentially univalent is equivalent with the fact that the restriction to G of the first coordinate projection, $(x, y) \rightsquigarrow x$, is injective.*

Let us assume that u_0 is essentially univalent. We denote by $\tilde{E}_0$ the image of G under the mapping $(x, y) \rightsquigarrow x$. Thus $\tilde{E}_0$ is a vector subspace of E and $(x, y) \rightsquigarrow x$ is a bijection of G onto it. We denote by $\tilde{u}_0$ the compose of the sequence $x \rightsquigarrow (x, y) \rightsquigarrow y$ to be understood as a sequence of mappings $\tilde{E}_0 \to G \to F$. As $G_0 \subset G$, we have $\varrho(E_0) \subset \tilde{E}_0$; and Eq. (13.2) is satisfied.

Definition 13.2. – *Assume that u_0 is essentially univalent; the linear mapping $\tilde{u}_0 : \tilde{E}_0 \to F$ will be called the canonical extension of u_0 with respect to the pair (ϱ, σ).* *)

We have $u_0 = \tilde{u}_0$ if and only if $G = G_0$, i.e., if and only if the graph of the mapping $\varrho(x_0) \rightsquigarrow \sigma(u_0(x_0))$ is closed. For the applications which we have in mind, this is the case which interests us least.

We continue to assume that u_0 is essentially univalent. Using the bijection $(x, y) \rightsquigarrow x$ of G onto $\tilde{E}_0$, we may (and we shall) transfer the topology of G onto $\tilde{E}_0$ (the topology of G is the one induced by the product topology on $E \times F$). We shall refer to the topology thus defined on E_0 as the *graph topology*. Of course, the graph topology is locally convex and Hausdorff. Furthermore, $\tilde{E}_0$ is complete whenever both E and F are complete. The graph topology on $\tilde{E}_0$ is the coarsest locally convex topology such that the natural injection of $\tilde{E}_0$ into E, which we shall denote by i, and the mapping $\tilde{u}_0 : \tilde{E}_0 \to F$ are continuous.

Since G is the closure of G_0 in $E \times F$, $\varrho(E_0)$ is a dense (for the graph topology) linear subspace of $\tilde{E}_0$.

---

*) Most of the time, the mention "with respect to the pair (ϱ, σ)" will be omitted.

We close this section with a few words about the spectrum of $\tilde{E}_0$ when the latter carries the graph topology. Let $\tilde{\mathfrak{r}}$ be a continuous seminorm on $\tilde{E}_0$. There is a unique continuous seminorm $\mathfrak{r}$ on G (p. 38) such that, for all $\tilde{x} \in \tilde{E}_0$,

$$\tilde{\mathfrak{r}}(\tilde{x}) = \mathfrak{r}(i(\tilde{x}), \tilde{u}_0(\tilde{x})).$$

As G carries the topology induced by the product TVS $E \times F$, there is (Lemma 1.1) a continuous seminorm on $E \times F$ extending $\mathfrak{r}$; we denote also by $\mathfrak{r}$ this extension. For $x \in E$ and $y \in F$, we set

$$\mathfrak{p}(x) = \mathfrak{r}(x, 0), \quad \mathfrak{q}(y) = \mathfrak{r}(0, y). \tag{13.4}$$

We have $\mathfrak{p} \in \operatorname{Spec} E$, $\mathfrak{q} \in \operatorname{Spec} F$ and

$$\mathfrak{r}(x, y) \leqslant \mathfrak{p}(x) + \mathfrak{q}(y).$$

Replacing x by $i(\tilde{x})$ and y by $\tilde{u}_0(\tilde{x})$ with $\tilde{x} \in \tilde{E}_0$ arbitrary, we see that

$$\tilde{\mathfrak{r}} \leqslant i_* \mathfrak{p} + \tilde{u}_{0*} \mathfrak{q}. \tag{13.5}$$

One might say that the irreducible hull of $(\operatorname{Im} i_*) \cup (\operatorname{Im} \tilde{u}_{0*})$ is the whole spectrum of $\tilde{E}_0$.

14. Existence and approximation of solutions

We keep the set-up and the notation of the previous section, 13, in particular, of Diagram (13.1). We suppose throughout the present section that u_0 is essentially univalent with respect to the pair (ϱ, σ) (Definition 13.1). We shall deal with the canonical extension (Definition 13.2) of u_0,

$$\tilde{u}_0 : \tilde{E}_0 \to F.$$

The space $\tilde{E}_0 \subset E$ carries the graph topology (p. 38).

Our purpose, here, is to establish that certain conditions, bearing on $u_{0*} : \operatorname{Spec} F_0 \to \operatorname{Spec} E_0$ (and involving ϱ and σ), imply certain interesting properties of existence and approximation of solutions to the equation

$$\tilde{u}_0(x) = y.$$

We begin by describing these properties and, first of all, the one which is relative to the *homogeneous* equation.

Definition 14.1. – *We say that $\tilde{u}_0$ has the homogeneous approximation property if $\varrho(\operatorname{Ker} u_0)$ is dense in* $\operatorname{Ker} \tilde{u}_0$.

In this statement, "dense" is to be understood in the sense of the graph topology. But, as a matter of fact, in this case, as we are dealing with solutions of the homogeneous equation

$$\tilde{u}_0(x) = 0,$$

it coincides with "dense in the sense of the topology induced by E''. The homogeneous approximation property can be restated without any reference to the canonical extension of u_0; it says the following:

Every $x \in E$ which is the limit of elements $\varrho(x_0)$ where the $x_0 \in E_0$ are such that $u_0(x_0)$ converges to O in F, is also the limit of elements $\varrho(\bar{x}_0)$, where the $\bar{x}_0 \in E_0$ verify $u_0(\bar{x}_0) = O$. (14.1)

The second property in which we shall be interested concerns the *inhomogeneous* equation. It simply says that the canonical extension $\tilde{u}_0$ is an epimorphism of $\tilde{E}_0$ onto F – property which can be stated as follows, without reference to the canonical extension:

To every neighborhood U of O in E, there is a neighborhood V of O in F such that every element $y \in V$ is the limit, in F, of elements $u_0(x_0)$, where the $x_0 \in E_0$ are such that $\varrho(x_0)$ converges, in E, to some element of U. (14.2)

If M is a vector subspace of E_0, we recall that M^0 is the set of seminorms $\mathfrak{p}_0 \in \operatorname{Spec} E_0$ vanishing on M. The criterion that we are seeking, implying the validity of (14.1) and (14.2), will be an immediate consequence of the following more general result:

Theorem 14.1. – *Suppose that the continuous linear map $u_0: E_0 \to F_0$ is essentially univalent with respect to the pair of continuous linear mappings $\varrho: E_0 \to E$, $\sigma: F_0 \to F$. Then the following properties are equivalent:*

(a) *the canonical extension $\tilde{u}_0$ of u_0 is a homomorphism and has the homogeneous approximation property;*

(b) $$(\operatorname{Im} \varrho_*) \cap (\operatorname{Ker} u_0)^0 \subset u_{0*}(\operatorname{Im} \sigma_*).$$

For the meaning of "essentially univalent" see Definition 13.1, for "canonical extension", Definition 13.2, for "homogeneous approximation property", Definition 14.1.

Proof: We begin by establishing a certain number of formulas. In these, we must distinguish carefully between $\varrho: E_0 \to E$ and the "same" mapping but regarded as taking its values in $\tilde{E}_0$; we shall denote the latter by $\tilde{\varrho}$. We recall that $i: \tilde{E}_0 \to E$ is the natural injection. We have

$$\varrho = i \circ \tilde{\varrho}, \quad \tilde{u}_0 \circ \tilde{\varrho} = \sigma \circ u_0.$$

Therefore:

$$\varrho_* = \tilde{\varrho}_* \circ i_*, \tag{14.3}$$

$$\tilde{\varrho}_* \circ \tilde{u}_{0*} = u_{0*} \circ \sigma_*. \tag{14.4}$$

In view of (14.3), we have

$$\operatorname{Im} \rho_* \subset \operatorname{Im} \tilde{\varrho}_*. \tag{14.5}$$

In view of (14.4),

$$\tilde{\varrho}_*(\operatorname{Im} \tilde{u}_{0*}) = u_{0*}(\operatorname{Im} \sigma_*), \tag{14.6}$$

whence, also,

$$u_{0*}(\operatorname{Im} \sigma_*) \subset (\operatorname{Im} \tilde{\varrho}_*) \cap (\operatorname{Ker} u_0)^0. \tag{14.7}$$

We shall prove the following fact:

(*) *the irreducible hull* (p. 4) *of*

$$(u_{0*}(\operatorname{Im} \sigma_*)) \cup [(\operatorname{Im} \varrho_*) \cap (\operatorname{Ker} u_0)^0] \tag{14.8}$$

is equal to

$$(\operatorname{Im} \tilde{\varrho}_*) \cap (\operatorname{Ker} u_0)^0. \tag{14.9}$$

Proof of (*)

In virtue of (14.5) and (14.7), the set (14.8) is contained in (14.9). It is the inclusion the other way around that we must prove.

Let $\tilde{\mathfrak{r}}$ be a continuous seminorm on $\tilde{E}_0$ (for the graph topology) and let $\mathfrak{p} \in \operatorname{Spec} E$, $\mathfrak{q} \in \operatorname{Spec} F$ be associated with it as in (13.4). It is clear that if $\tilde{\mathfrak{r}}=0$ on $\tilde{\varrho}(\operatorname{Ker} u_0)$, then $\mathfrak{p}=0$ on $\varrho(\operatorname{Ker} u_0)$. We suppose this to be the case. Consider then the seminorm on $\tilde{E}_0$

$$\tilde{\mathfrak{r}}_1 = i_*\mathfrak{p} + \tilde{u}_{0*}\mathfrak{q}.$$

We have

$$\tilde{\varrho}_*\tilde{\mathfrak{r}}_1 = \varrho_*\mathfrak{p} + u_{0*}(\sigma_*\mathfrak{q})$$

in view of (14.3) and (14.4). As $\mathfrak{p} \in (\varrho(\operatorname{Ker} u_0))^0$, we have $\varrho_*\mathfrak{p} \in (\operatorname{Im} \varrho_*) \cap (\operatorname{Ker} u_0)^0$, whence the fact that $\tilde{\varrho}_*\tilde{\mathfrak{r}}_1$ and, a fortiori, $\tilde{\varrho}_*\tilde{\mathfrak{r}}$ belong to the set (14.8). This proves (*).

Now, Property (b) in Theorem 14.1 is equivalent with the fact that the irreducible hull of (14.8) is contained in $u_{0*}(\operatorname{Im} \sigma_*)$. In view of (14.7), we see that (b) is equivalent with the property:

$$u_{0*}(\operatorname{Im} \sigma_*) = (\operatorname{Im} \tilde{\varrho}_*) \cap (\operatorname{Ker} u_0)^0.$$

This, in turn, is equivalent, because of (14.6), with

(b′) $$\tilde{\varrho}_*(\operatorname{Im} \tilde{u}_{0*}) = (\operatorname{Im} \tilde{\varrho}_*) \cap (\operatorname{Ker} u_0)^0.$$

We use now the obvious fact that

$$\tilde{\varrho}_*[(\tilde{\varrho}(\operatorname{Ker} u_0))^0] = (\operatorname{Im} \tilde{\varrho}_*) \cap (\operatorname{Ker} u_0)^0.$$

It shows that (b′) is equivalent with

$$\tilde{\varrho}_*(\operatorname{Im}\tilde{u}_{0*}) = \tilde{\varrho}_*[(\tilde{\varrho}(\operatorname{Ker} u_0))^0],$$

but as $\operatorname{Im}\tilde{\varrho}$ is dense, hence $\tilde{\varrho}_*$ is injective, we conclude that (b′), hence (b), is equivalent with

(b″) $$\operatorname{Im}\tilde{u}_{0*} = (\tilde{\varrho}(\operatorname{Ker} u_0))^0.$$

Property (b″) in turn is equivalent with the conjunction of the following two properties:

(b_1'') $$\operatorname{Im}\tilde{u}_{0*} = (\operatorname{Ker}\tilde{u}_0)^0;$$

(b_2'') $$(\tilde{\varrho}(\operatorname{Ker} u_0))^0 = (\operatorname{Ker}\tilde{u}_0)^0.$$

Indeed, we have trivially, in any case,

$$\operatorname{Im} u_{0*} \subset (\operatorname{Ker}\tilde{u}_0)^0,$$

$$(\operatorname{Ker}\tilde{u}_0)^0 \subset (\tilde{\varrho}(\operatorname{Ker} u_0))^0,$$

the latter being a consequence of the obvious fact that $\tilde{\varrho}(\operatorname{Ker} u_0) \subset \operatorname{Ker}\tilde{u}_0$.

Now, by Proposition 3.1, (b_2'') is equivalent with the fact that $\tilde{u}_0$ has the homogeneous approximation property, whereas, by Corollary 1 of Proposition 4.7, (b_1'') is equivalent with the fact that $\tilde{u}_0$ is a homomorphism of $\tilde{E}_0$ into F. Q.E.D.

We may now derive the sought criterion:

Corollary 1. – *Same hypotheses as in Theorem* 14.1. *Suppose moreover that the space F is complete. Then the following properties are equivalent*:

(a) $\tilde{u}_0$, *the canonical extension of* u_0, *is an epimorphism and has the homogeneous approximation property*;

(b) (b_1) $(\operatorname{Im}\varrho_*) \cap (\operatorname{Ker} u_0)^0 \subset u_{0*}(\operatorname{Im}\sigma_*)$;

(b_2) $u_{0*} \circ \sigma_*$ *is injective*;

(b_3) $\tilde{E}_0/\operatorname{Ker}\tilde{u}_0$ *is complete.*

Proof: Property (b_2) is equivalent with the fact that $\operatorname{Im}\tilde{u}_0$ is dense in F. In virtue of Theorem 14.1, (b_1) says that $\tilde{u}_0$ is a homomorphism and has the homogeneous approximation property. It remains to use the fact that, when F is complete, a homomorphism with dense image, $\tilde{u}_0 : \tilde{E}_0 \to F$, is surjective if and only if the quotient TVS $\tilde{E}_0/\operatorname{Ker}\tilde{u}_0$ is complete. And this is obvious.

Chapter V. Translation into Duality

15. The seminorms "absolute value of a linear functional"

Let E be a vector space over the field $\mathbf{K}=\mathbf{R}$ or $\mathbf{C}$. We denote by E^* the *algebraic dual* of E, i.e., the vector space of the linear mappings $E \to \mathbf{K}$. Elements of E^* will be denoted by x^*, x_1^*, y^*, etc.; the value of the functional x^* at the point $x \in E$ will be denoted by $\langle x^*, x\rangle$. We denote by $|x^*|$ the seminorm on E,

$$x \rightsquigarrow |\langle x^*, x\rangle| .$$

Most properties introduced in the earlier chapters, which are relative to seminorms, can be checked by looking only at the seminorms of the kind $|x^*|$, i.e., of the kind "absolute value of a linear functional". The reason for this lies in two very simple lemmas which are going to be stated and proved in the present section. The first of these lemmas embodies a substantial part of the Hahn-Banach theorem. The second one is essentially equivalent with Mackey's theorem stating that the bounded subsets of a locally convex TVS are the same for all the locally convex topologies giving rise to the same *continuous* dual (the continuous dual E' of E is the vector space of the continuous linear functionals on E).

We shall say that a seminorm $\mathfrak{p}$ on E is *extremal* if for every seminorm $\mathfrak{q}$ on E, $\mathfrak{q} \leqslant \mathfrak{p}$ implies that there is a constant $C \geqslant 0$ such that $\mathfrak{q} = C\mathfrak{p}$.

Proposition 15.1. – *The following properties of a seminorm* $\mathfrak{p}$ *on* E *are equivalent*:

(a) *there is a linear functional* $x^* \neq 0$ *on* E *such that* $\mathfrak{p} = |x^*|$;

(b) $\operatorname{Ker}\mathfrak{p}$ *is a hyperplane of* E;

(c) $\hat{E}_{\mathfrak{p}}$ *is one-dimensional*;

(d) $\hat{E}_{\mathfrak{p}}$ *is a one-dimensional Hilbert space*;

(e) $\mathfrak{p}$ *is extremal.*

Proof: (a)⇒(b)⇒(c) trivially. Suppose that (c) holds; then $\| \ \|_{\mathfrak{p}}$ is a norm on $\hat{E}_{\mathfrak{p}}$ in the sense of the field $\mathbf{K}$: but on a one-dimensional vector space, all norms are Hilbert norms, whence (d). Suppose that (d) holds and let $\omega_{\mathfrak{p}}$ be the canonical mapping of E into $\hat{E}_{\mathfrak{p}}$; there is a linear functional on $\hat{E}_{\mathfrak{p}}$, z^*, whose absolute value is equal to the norm $\| \ \|_{\mathfrak{p}}$.

Then $(z^*)\circ\omega_{\mathfrak{p}}$ is a linear functional on E whose absolute value is equal to $\mathfrak{p}$. Thus (a), (b), (c) and (d) are equivalent.

Suppose next that (b) holds and let $\mathfrak{q}$ be a seminorm on E, $\mathfrak{q}\leqslant\mathfrak{p}$. We want to prove that $\mathfrak{q}$ is proportional to $\mathfrak{p}$. This is evident if $\mathfrak{q}=0$. Otherwise, let $x_0\in E$ be such that $\mathfrak{q}(x_0)\neq 0$. We have $x_0\notin \operatorname{Ker}\mathfrak{p}$, hence every $x\in E$ can be written as $x=\lambda x_0+x_1$ with $\lambda\in\mathbf{K}$ and $x_1\in\operatorname{Ker}\mathfrak{p}\subset\operatorname{Ker}\mathfrak{q}$. Therefore,

$$\mathfrak{p}(x)=\mathfrak{p}(\lambda x_0)=|\lambda|\,\mathfrak{p}(x_0),\quad \mathfrak{q}(x)=\mathfrak{q}(\lambda x_0)=|\lambda|\,\mathfrak{q}(x_0),$$

whence $\mathfrak{p}(x)\,\mathfrak{q}(x_0)=\mathfrak{p}(x_0)\,\mathfrak{q}(x)$. This means that (e) holds.

Suppose that (b) does not hold. Let M be a vector subspace of E, distinct from E and from $\operatorname{Ker}\mathfrak{p}$, such that $\operatorname{Ker}\mathfrak{p}\subset M$. The seminorm

$$x\rightsquigarrow \inf_{x-y\in M}\mathfrak{p}(y)$$

is $\leqslant\mathfrak{p}$ but is not proportional to $\mathfrak{p}$. Thus (e) does not hold. Q.E.D.

Next, we state the first one of the two lemmas announced at the beginning:

Lemma 15.1. – *Let M be a vector subspace of E, $\mathfrak{p}$ a seminorm on E, x_0^* a linear functional on M such that $|\langle x_0^*, x\rangle|\leqslant\mathfrak{p}(x)$ for all $x\in M$. There is a linear functional x^* on E, extending x_0^* and such that $|x^*|\leqslant\mathfrak{p}$.*

This is one of the standard versions of the Hahn-Banach theorem and its proof will not be given here.

Corollary 1. – *Let $\mathfrak{p}$ be a seminorm on E, x a point of E. There is a linear functional x^* on E such that $|x^*|\leqslant\mathfrak{p}$ and $\langle x^*, x\rangle=\mathfrak{p}(x)$.*

Proof: Let M be the subspace of E spanned by x; we apply Lemma 15.1 to M, $\mathfrak{p}$ and to $x_0^*:\lambda x\rightsquigarrow\lambda\mathfrak{p}(x)$.

Corollary 2. – *Every seminorm is the upper enveloppe of the seminorms "absolute value of a linear functional" which are at most equal to it.*

Corollary 3. – *Let E be a locally convex TVS, A a closed convex balanced subset of E, x a point of E which does not belong to A. There is a continuous linear functional x' on E such that:*

$$|\langle x', x\rangle|>1,\quad \sup_{a\in A}|\langle x', a\rangle|\leqslant 1\,.$$

Proof: Combine Corollary 1 with Lemma 3.1.

In order to prove the second basic lemma, we recall a few well known facts about normed spaces. Let $(E,\mathfrak{p})$ be a normed space: E is the underlying vector space, $\mathfrak{p}$ the norm. Let E' be the dual of E; for $x'\in E'$ arbitrary, we set

$$\mathfrak{p}'(x')=\sup_{x\in E,\ \mathfrak{p}(x)=1}|\langle x', x\rangle|\,.$$

It is obvious that $\mathfrak{p}'$ is a norm on E' and it is easy to see (cf. proof of Proposition 6.4) that the normed space $(E', \mathfrak{p}')$ is a Banach space. Furthermore, we have, for all $x \in E$,

$$\mathfrak{p}(x) = \sup_{x' \in E',\ \mathfrak{p}'(x')=1} |\langle x', x \rangle| . \tag{15.1}$$

Proof of (15.1): Let $B_{\mathfrak{p}}$ (resp. $B_{\mathfrak{p}'}$) denote the closed unit ball of the norm $\mathfrak{p}$ (resp. $\mathfrak{p}'$), and let $B^0_{\mathfrak{p}'}$ denote the subset of E,

$$\{x \in E; \sup_{x' \in E',\ \mathfrak{p}'(x')=1} |\langle x', x \rangle| \leqslant 1\} .$$

Obviously, $B^0_{\mathfrak{p}'}$ is closed, convex, balanced and contains $B_{\mathfrak{p}}$. If $B_{\mathfrak{p}} \neq B^0_{\mathfrak{p}'}$ there would be a point $x \in B^0_{\mathfrak{p}'}$, $x \notin B_{\mathfrak{p}}$ and a continuous linear functional x' on E such that $|\langle x', x \rangle| > 1$ and $\sup_{y \in B_{\mathfrak{p}}} |\langle x', y \rangle| \leqslant 1$ (Corollary 3 of Lemma 15.1). The first inequality implies that $x' \notin B^0_{\mathfrak{p}}$ whereas the second one means exactly the contrary. Q.E.D.

We may now prove Mackey's lemma:

Lemma 15.2. – *Let $\mathfrak{p}$ be a seminorm on E, A a subset of E. The following conditions are equivalent:*

(a) $\mathfrak{p}$ *is bounded on A, i.e.,* $\sup_{x \in A} \mathfrak{p}(x) < +\infty$;

(b) *for all $x' \in E'$ such that* $|x'| \leqslant \mathfrak{p}$, $\sup_{x \in A} |\langle x', x \rangle| < +\infty$.

Proof: Nothing is changed if we replace E by $\hat{E}_{\mathfrak{p}}$ and A by its image under the canonical map $\omega_p : E \to \hat{E}_{\mathfrak{p}}$. Notice that the $x' \in E'$ such that $|x'| \leqslant \mathfrak{p}$ make up exactly the image under ${}^t\omega_{\mathfrak{p}}$ of the closed unit ball of $(\hat{E}_{\mathfrak{p}})'$. In other words, we may assume that we are dealing with a normed space, namely $(E, \mathfrak{p})$.

We suppose that (b) holds. Consider

$$T' = \bigcap_{x \in A} \{x' \in E'; |\langle x', x \rangle| \leqslant 1\} .$$

The subset T' of E' is closed convex balanced; in view of (b), T' is absorbent: indeed, observe that every $x' \in E'$ satisfies now $|x'| < C\mathfrak{p}$ for some $C > 0$. Thus T' is a barrel in E', hence a neighborhood of zero as E' is a Banach space (corollary of Proposition 9.3). This means that if $\mathfrak{p}'$ is the dual norm on E', we have, for some $c > 0$, $cB_{\mathfrak{p}'} \subset T'$, which implies, in virtue of (15.1):

$$T'^0 = \{x \in E; \sup_{x' \in T'} |\langle x', x \rangle| \leqslant 1\} \subset c^{-1} B_{\mathfrak{p}} .$$

But T'^0 contains obviously A, hence $\mathfrak{p}(x) \leqslant c^{-1}$ for all $x \in A$. Q.E.D.

Locally convex topologies compatible with a duality

Let F be a second vector space, over the same scalar field as E. We suppose that we are given a *duality bracket* $\langle\, ,\rangle$ on $E\times F$, i.e., a bilinear functional on this product space. For every $y\in F$, we may consider the linear functional on E, $x \rightsquigarrow \langle x, y\rangle$; we shall denote by $|y|$ its absolute value; it is a seminorm on E: $|y|(x)=|\langle x,y\rangle|$. We assume that the seminorm $|y|$ vanishes identically in E only if $y=0$. Let then $\sigma(F)$ be the irreducible hull (p. 4) of the set of seminorms $|y|$ as y varies over F. It is not difficult, now, to see what is meant by a *locally convex topology on E compatible with the duality between E and F*: it is the data of an irreducible subset S of Spec E such that $\sigma(F)$ is exactly the irreducible hull of the set of extremal seminorms which belong to S. In other words, it is an irreducible set $S\in$ Spec E such that if a seminorm $|x^*|$, $x^*\in E^*$, belongs to S, then there is an element y of F such that $|y|=|x^*|$. In view of Lemma 15.1, this means that the dual of the locally convex space E, equipped with the topology defined by S, can be identified with F. The irreducible sets $S\subset$ Spec E having this property, hence the locally convex topologies on E compatible with the duality between E and F, form a lattice, whose minimum element is evidently $\sigma(F)$ (this defines the *weak* topology $\sigma(E, F)$), and whose maximum element is the irreducible hull $\tau(F)$ of the set of seminorms $\mathfrak{p}$ on E of the kind

$$\mathfrak{p} = \sup_{y\in K} |y|\,,$$

where K is a convex balanced subset of F having the following property:

$$\begin{array}{l}\textit{for every } x^*\in E^*, \textit{ if, for all } x\in E, \textit{ the number } \langle x^*, x\rangle \\ \textit{belongs to the closure of the set of scalars } \langle x, y\rangle, \\ y\in K, \textit{ then } x^* \textit{ must be of the form } x\rightsquigarrow\langle x, y_0\rangle,\ y_0\in K.\end{array} \tag{15.2}$$

Property (15.2) is often stated by saying that K is *weakly compact*, or compact for the weak topology $\sigma(F, E)$. There are, in what precedes, a number of assertions that must be checked: 1°) that every balanced convex subset K of F which is contained in a finite dimensional vector subspace of F, and closed there (for the topology $\sigma(F, E)$ – or in the usual sense if $\sigma(F, E)$ is Hausdorff) verifies (15.2); 2°) that every extremal seminorm belonging to $\tau(F)$ is of the form $|y|$, $y\in F$ (this is practically evident by using (15.2)); 3°) that $\tau(F)$ is indeed maximum among the irreducible subsets of Spec E having the property in question. These statements are well known, and their proof will not be given. The topology on E defined by $\tau(F)$ is called *Mackey's topology* and often denoted by $\tau(E, F)$.

A subset B of a locally convex space E is said to be *bounded* if every continuous seminorm $\mathfrak{p}$ on E is bounded on B. It follows at once from Lemma 15.2 that a set $B \subset E$ which is bounded for the topology $\sigma(E, F)$ is also bounded for $\tau(E, F)$. In other words, and this is a standard statement of Mackey's theorem, alias Lemma 15.2, *the bounded sets are the same for all locally convex topologies compatible with a given duality*: indeed, a set is bounded if and only if every seminorm $|x'|$, $x' \in E'$, is bounded on it: its boundedness depends only on the extremal seminorms which are contained in $\operatorname{Spec} E$.

16. Lower star and transpose

We consider two locally convex Hausdorff TVS, E and F, and a continuous linear map

$$u: E \to F.$$

We recall that the transpose of u is the mapping

$${}^t u: F' \to E',$$

defined by

$$\langle y', u(x)\rangle = \langle {}^t u(y'), x\rangle, \qquad x \in E, \quad y' \in F'.$$

If we take the absolute values of both sides of the preceding identity, we obtain

$$u_*(|y'|) = |{}^t u(y')|, \qquad y' \in F'. \tag{16.1}$$

Proposition 16.1. – *The continuous linear functional on E, x', belongs to* $\operatorname{Im} {}^t u$ *if and only if the seminorm $|x'|$ belongs to* $\operatorname{Im} u_*$.

Proof: In virtue of (16.1), $x' \in \operatorname{Im} {}^t u$ implies $|x'| \in \operatorname{Im} u_*$. Conversely, suppose that $|x'| = u_* \mathfrak{q}$ for some $\mathfrak{q} \in \operatorname{Spec} F$. Then $u(x) \rightsquigarrow \langle x', x\rangle$ is a linear functional on $\operatorname{Im} u$, whose absolute value is $\leqslant \mathfrak{q}$. In view of the Hahn-Banach theorem (Lemma 15.1), it can be extended as a continuous linear functional y' on F; we have $x' = {}^t u(y')$. Q.E.D.

We present now a few further consequences of Lemmas 15.1 and 15.2.

Proposition 16.2. – *The following conditions are equivalent*:

(a) $\operatorname{Im} u$ *is dense in* F;

(b) $u_*: \operatorname{Spec} F \to \operatorname{Spec} E$ *is injective*;

(c) ${}^t u: F' \to E'$ *is injective.*

Proposition 16.2 is an immediate consequence of Proposition 4.2 and Corollary 2 of Lemma 15.1.

We recall that a subset B of the locally convex space F is said to be *bounded* if every continuous seminorm on F is bounded on B.

Proposition 16.3. – *Suppose that the mapping u has the following properties:*

the closure of $\mathrm{Im}\,u$ *in F,* $\overline{\mathrm{Im}\,u}$, *is a barrelled space* (Definition 9.2); (16.2)

every element y of $\overline{\mathrm{Im}\,u}$ *belongs to the closure of a subset of* $\mathrm{Im}\,u$ *which is bounded in F.* (16.3)

Then, an arbitrary seminorm $\mathfrak{p}\in\mathrm{Spec}\,E$ *belongs to* $\mathrm{Im}\,u_*$ *if and only if every linear functional x' on E such that* $|x'|\leqslant\mathfrak{p}$ *belongs to* $\mathrm{Im}\,{}^tu$.

Proof: If $\mathfrak{p}\in\mathrm{Im}\,u_*$ we have $|x'|\in\mathrm{Im}\,u_*$ for all $x'\in E'$, $|x'|\leqslant\mathfrak{p}$, as $\mathrm{Im}\,u_*$ is an irreducible set (Proposition 6.1), therefore $x'\in\mathrm{Im}\,{}^tu$ in view of Proposition 16.1.

We assume now that Properties (16.2) & (16.3) hold and that every $x'\in E'$, $|x'|\leqslant\mathfrak{p}$, is equal to ${}^tu(y')$ for some $y'\in F'$. Let $y\in\overline{\mathrm{Im}\,u}$ be arbitrary. There is a subset A of E such that $y\in\overline{u(A)}$ and such that $u(A)$ is bounded in F. Given $x'\in E'$, $|x'|\leqslant\mathfrak{p}$, and $y'\in F'$, ${}^tu(y')=x'$, there is a constant $C>0$ such that, for all $x\in A$,

$$|\langle y', u(x)\rangle| = |\langle x', x\rangle| \leqslant C.$$

This proves that every seminorm of the type $|x'|\leqslant\mathfrak{p}$ is bounded on A. It follows then, from Lemma 15.2, that $\mathfrak{p}$ itself is bounded on A. Therefore we have, for all $y'\in F'$ such that $|{}^tu(y')|\leqslant\mathfrak{p}$,

$$|\langle y', y\rangle| \leqslant \sup_{x\in A} \mathfrak{p}(x).$$

This means that, for every $y\in\overline{\mathrm{Im}\,u}$,

$$\sup_{y'\in F',\ |{}^tu(y')|\leqslant\mathfrak{p}} |\langle y', y\rangle| < +\infty.$$

The upper enveloppe of the seminorms $|y'|$, $|{}^tu(y')|\leqslant\mathfrak{p}$, is a continuous seminorm $\mathfrak{q}$ on $\overline{\mathrm{Im}\,u}$, in view of (16.2) and of Proposition 9.2; it can be extended as a continuous seminorm on F, which we denote also by $\mathfrak{q}$ (Lemma 1.1). As a last step, we show that $\mathfrak{p}=u_*\mathfrak{q}$. Indeed, let x be an arbitrary point of E; in view of Corollary 1 of Lemma 15.1, there is $x'\in E'$, $|x'|\leqslant\mathfrak{p}$, such that $\langle x', x\rangle=\mathfrak{p}(x)$. By hypothesis, there is $y'\in F'$ such that $x'={}^tu(y')$ and we have $|y'|\leqslant\mathfrak{q}$, therefore $\mathfrak{p}(x)=\langle x', x\rangle=\langle y', u(x)\rangle\leqslant\mathfrak{q}(u(x))$. Conversely, given $\varepsilon>0$ arbitrary, there is $y'\in F'$ such that $|{}^tu(y')|\leqslant\mathfrak{p}$ and such that $|y'|\leqslant\mathfrak{q}$ and $\mathfrak{q}(u(x))\leqslant|\langle y', u(x)\rangle|+\varepsilon$. As the latter quantity is $\leqslant\mathfrak{p}(x)+\varepsilon$, and as ε is arbitrary, we conclude that $\mathfrak{q}(u(x))\leqslant\mathfrak{p}(x)$. Q.E.D.

Corollary 1. – *Suppose that F is a Fréchet space. Then* $\mathfrak{p}\in\mathrm{Spec}\,E$ *belongs to* $\mathrm{Im}\,u_*$ *if and only if* $x'\in\mathrm{Im}\,{}^tu$ *for all* $x'\in E'$ *such that* $|x'|\leqslant\mathfrak{p}$.

Indeed, every closed vector subspace of F, in particular $\overline{\operatorname{Im} u}$, is barrelled; and a point y belongs to the closure of a subset M of F if and only if y belongs to the closure of a convergent sequence of F contained in M, whence the corollary.

Corollary 2. – *Under the same hypotheses as in Proposition* 16.3, (16.2) & (16.3), *the following conditions are equivalent*:

(a) *u is a homomorphism*;

(b) $\operatorname{Im} {}^t u = (\operatorname{Ker} u)^\perp$.

If M is a vector subspace of E, we denote by $M^\perp$ the *orthogonal* of M, i.e., the vector subspace of E' consisting of the continuous linear functionals which vanish on M.

Proof of Corollary 2: We apply Corollary 1 of Proposition 4.7, in view of which (a) is equivalent with the fact that $\operatorname{Im} u_* = (\operatorname{Ker} u)^0$. Now, observe that, for all $x' \in E'$, $x' \in (\operatorname{Ker} u)^\perp$ is equivalent with $|x'| \in (\operatorname{Ker} u)^0$. Suppose then that (a) holds and that $x' \in (\operatorname{Ker} u)^\perp$, i.e., $|x'| \in (\operatorname{Ker} u)^0$, hence $|x'| \in \operatorname{Im} u_*$ hence, by Proposition 16.1, $x' \in \operatorname{Im} {}^t u$. Conversely, suppose that (b) holds and let $\mathfrak{p} \in (\operatorname{Ker} u)^0$. Then every $x' \in E'$ which satisfies $|x'| \leqslant \mathfrak{p}$ vanishes on $\operatorname{Ker} u$, hence belongs to $\operatorname{Im} {}^t u$. In view of Proposition 16.3, $\mathfrak{p}$ must belong to $\operatorname{Im} u_*$. Q.E.D.

Corollary 3. – *Same hypotheses as in Proposition* 16.3. *Then the following properties are equivalent*:

(a) *u is a homomorphism with dense image*;

(b) *${}^t u$ is injective and* $\operatorname{Im} {}^t u = (\operatorname{Ker} u)^\perp$.

Combine Corollary 2 with Proposition 16.2.

Corollary 4. – *Same hypotheses as in Proposition* 16.3. *Then the following properties are equivalent*:

(a) *u is a monomorphism*;

(b) *${}^t u$ is surjective.*

As a monomorphism is an injective homomorphism, Corollary 4 follows at once from Corollary 2.

Corollary 5. – *Same hypotheses as in Proposition* 16.3. *Then the following properties are equivalent*:

(a) *u is an isomorphism*;

(b) *${}^t u$ is bijective.*

Let us observe that the fact that

$$\operatorname{Im} {}^t u = (\operatorname{Ker} u)^\perp$$

is equivalent with the fact that $\operatorname{Im} {}^t u$ is *weakly closed* (the weak dual topology on E' is simply the topology of pointwise convergence of continuous linear functionals). Indeed, it is easy to see that $\operatorname{Im}{}^t u$ is weakly dense in $(\operatorname{Ker} u)^\perp$; and it is obvious that $(\operatorname{Ker} u)^\perp$ is weakly closed.

We go back to the general situation of two locally convex Hausdorff TVS, E and F, and of a continuous linear map $u: E \to F$ (in other words, we do not assume any more that (16.2) & (16.3) hold).

Proposition 16.4. – *Let* $\mathfrak{p} \in \operatorname{Spec} E$, $\mathfrak{q} \in \operatorname{Spec} F$. *The following conditions are equivalent:*

(a) *for every* $\mathfrak{r} \in \operatorname{Spec} F$, $u_* \mathfrak{r} \leqslant \mathfrak{p}$ *implies* $\mathfrak{r} \leqslant \mathfrak{q}$;

(b) *for every* $y' \in F'$, $|{}^t u(y')| \leqslant \mathfrak{p}$ *implies* $|y'| \leqslant \mathfrak{q}$.

Follows at once from Corollary 2 of Lemma 15.1.

A subset A' of E' is *equicontinuous* if and only if the set of seminorms $|x'|$, $x' \in A'$, is equicontinuous (see Section 8).

Corollary 1. – *The following properties are equivalent:*

(a) *the preimage, under* u_*, *of any equicontinuous subset of* $\operatorname{Spec} E$ *is an equicontinuous subset of* $\operatorname{Spec} F$;

(b) *the preimage, under* ${}^t u$, *of any equicontinuous subset of* E' *is an equicontinuous subset of* F'.

Combine Propositions 8.1 and 16.4.

The dual of the Banach space $\hat{E}_\mathfrak{p}$

Let $\mathfrak{p}$ be a continuous seminorm on E, $\omega_\mathfrak{p}: E \to \hat{E}_\mathfrak{p}$ the canonical map, ${}^t\omega_\mathfrak{p}: (\hat{E}_\mathfrak{p})' \to E'$ its transpose. Since $\operatorname{Im} \omega_\mathfrak{p}$ is dense, ${}^t\omega_\mathfrak{p}$ is injective, and $(\hat{E}_\mathfrak{p})'$ can be identified, via ${}^t\omega_\mathfrak{p}$, with a vector subspace of E' (in fact, with a *Banach subspace* of E', i.e., a vector subspace equipped with a Banach space structure for which the injection into E', equipped, say, with the strong dual topology, is continuous). The image of ${}^t\omega_\mathfrak{p}$ is not hard to characterize: indeed, we know that $\omega_{\mathfrak{p}*}$ is a bijection of $\operatorname{Spec} \hat{E}_\mathfrak{p}$ onto the irreducible subset $S_\mathfrak{p}$ of $\operatorname{Spec} E$ (see pp. 22, 23): $S_\mathfrak{p}$ is the set of seminorms $\mathfrak{q}$ on E such that $\mathfrak{q} \leqslant C\mathfrak{p}$ for some constant $C \geqslant 0$. Therefore, ${}^t\omega_\mathfrak{p}$ is a bijection of $(\hat{E}_\mathfrak{p})'$ onto the vector subspace of E' consisting of the linear functionals x' such that $|x'| \in S_\mathfrak{p}$, i.e., which are continuous on E for the topology defined by the single seminorm p. Furthermore, the norm on $(\hat{E}_\mathfrak{p})'$ is exactly the norm associated with the metric $\|\mathfrak{q}_1 - \mathfrak{q}_2\|_\mathfrak{p}$ defined on the set $S_\mathfrak{p}$ (see p. 23):

$$\|x'\|_\mathfrak{p} = \sup_{x \in E,\ \mathfrak{p}(x) \neq 0} \left(\mathfrak{p}(x)^{-1} |\langle x', x\rangle|\right).$$

From now on, we identify $(\hat{E}_\mathfrak{p})'$ with its image under ${}^t\omega_\mathfrak{p}$.

17. Duality in Relation with Existence and Approximation of Solutions to a Functional Equation

We continue to consider two locally convex Hausdorff TVS, E and F, and a continuous linear map $u: E \to F$. In this section we shall translate, into duality terms, the existence and approximation theorems of Chapters III & IV.

We begin with Theorem 10.1:

Theorem 17.1. – *Suppose that E and F are Fréchet spaces. The following properties are equivalent:*

(a) *u is an epimorphism;*

(b) *u is surjective;*

(c) *u is a homomorphism with dense image;*

(d) *u is presurjective;*

(e) *the preimage, under ${}^t u$, of any equicontinuous subset of E' is an equicontinuous subset of F';*

(f) *${}^t u$ is injective and* $\mathrm{Im}\,{}^t u = (\mathrm{Ker}\, u)^{\perp}$.

It suffices to combine Theorem 10.1 with Corollary 3 of Proposition 16.3 and with Corollary 1 of Proposition 16.4 (note that Fréchet spaces have always Properties (16.2) & (16.3) in Proposition 16.3: cf. proof of Corollary 1 of the same).

We come next to the translation into duality of some of the criteria of presurjectivity of Section 11. We begin by studying Proposition 11.1. If we make use of the remarks on p. 50, (see also p. 23), we may state the following:

Proposition 17.1. – *Suppose that there is a basis of continuous seminorms on E (resp. on F), $\mathfrak{B}$ (resp. $\mathfrak{B}'$) such that the following is true:*

To every $\mathfrak{p} \in \mathfrak{B}$ there is $\mathfrak{q} \in \mathfrak{B}'$ and a constant $C > 0$ such that, for all $y' \in F'$, the fact that ${}^t u(y') \in (\hat{E}_{\mathfrak{p}})'$ implies that $y' \in (\hat{F}_{\mathfrak{q}})'$ and that (17.1)

$$\|y'\|_{\mathfrak{q}} \leqslant C\, \|{}^t u(y')\|_{\mathfrak{p}}. \tag{17.2}$$

Then u is presurjective. Conversely, if F is barrelled and if u is presurjective, then (17.1) *holds for any pair $\mathfrak{B}$, $\mathfrak{B}'$ of bases of continuous seminorms on E and F respectively.*

It suffices to combine Propositions 11.1 and 16.4.

We must also mention the following immediate consequence of Corollary 2 of Lemma 15.1:

Proposition 17.2. – *The pair* (E, F) *is u-convex* (Definition 11.1) *if and only if, to every seminorm* $\mathfrak{p} \in \operatorname{Spec} E$ *there is a seminorm* $\mathfrak{q} \in \operatorname{Spec} F$ *such that, for all* $y' \in F'$,

$$ {}^t u(y') \in (\hat{E}_{\mathfrak{p}})' \quad \textit{implies} \quad y' \in (\operatorname{Ker} \mathfrak{q})^{\perp}. $$

We switch now to the existence and approximation results of Chapter IV. We consider, as we did there, the diagram (13.1):

$$ \begin{array}{ccc} E_0 & \xrightarrow{\varrho} & E \\ {\scriptstyle u_0} \downarrow & & \\ F_0 & \xrightarrow[\sigma]{} & F. \end{array} $$

All the four spaces under consideration are locally convex and Hausdorff; the three mappings u_0, ϱ, σ are linear and continuous.

Proposition 17.3. – *Suppose that the following conditions are satisfied*:

$\overline{\sigma(\operatorname{Im} u_0)}$, *equipped with the topology induced by* F, *is a barrelled space*; (17.3)

every element y *of* $\overline{\sigma(\operatorname{Im} u_0)}$ *belongs to the closure of a subset of* $\sigma(\operatorname{Im} u_0)$ *which is bounded in* F. (17.4)

Then the following properties are equivalent:

(a) $$ (\operatorname{Im} \varrho_*) \cap (\operatorname{Ker} u_0)^0 \subset u_{0*}(\operatorname{Im} \sigma_*); $$

(b) *for every* $x' \in E'$,

$$ {}^t\varrho(x') \in (\operatorname{Ker} u_0)^{\perp} \quad \textit{implies} \quad {}^t\varrho(x') \in {}^t u_0(\operatorname{Im} {}^t\sigma). $$

Proof: Let us set $v = \sigma \circ u_0$, whence $v_* = u_{0*} \circ \sigma_*$; (a) reads then $(\operatorname{Im} \varrho_*) \cap (\operatorname{Ker} u_0)^0 \subset \operatorname{Im} v_*$. But also ${}^t v = {}^t u_0 \circ {}^t\sigma$, so that (b) reads: ${}^t\varrho(x') \in (\operatorname{Ker} u_0)^{\perp}$ *implies* ${}^t\varrho(x') \in \operatorname{Im} {}^t v$ *whatever* $x' \in E'$. Proposition 16.1 implies that (a)$\Rightarrow$(b). Conversely, let $\mathfrak{p}_0 \in (\operatorname{Im} \varrho)_* \cap (\operatorname{Ker} u_0)^0$; then every $|x_0'| \leqslant \mathfrak{p}_0$ $(x_0' \in E_0')$ belongs to $\operatorname{Im} \varrho_*$, hence (Proposition 16.1) to $\operatorname{Im} {}^t\varrho$, and obviously also to $(\operatorname{Ker} u_0)^{\perp}$. If (b) holds, $|x_0'| \in \operatorname{Im} {}^t v$. In view of Hypotheses (17.3) and (17.4), we may apply Proposition 16.3; we reach thus the conclusion that $\mathfrak{p}_0 \in \operatorname{Im} v_*$.

Q.E.D.

We may now translate Theorem 14.1. For this, we assume that u_0 is essentially univalent with respect to the pair (ϱ, σ) (Definition 13.1) and we denote by $\tilde{u}_0 : \tilde{E}_0 \to F$ the canonical extension (Definition 13.2) of u_0.

Theorem 17.2. – *We assume that the mapping* $u_0 : E_0 \to F_0$ *is essentially univalent with respect to the pair of mappings*

$$ \varrho : E_0 \to E, \quad \sigma : F_0 \to F. $$

We assume also that Conditions (17.3) & (17.4) *are fulfilled. Then the following properties are equivalent*:

(a) *the canonical extension* $\tilde{u}_0$ *of* u_0 *is a homomorphism and has the homogeneous approximation property* (Definition 14.1);

(b) *for every* $x' \in E'$,

$$ {}^t\varrho(x') \in (\operatorname{Ker} u_0)^\perp \quad \textit{implies} \quad {}^t\varrho(x') \in {}^t u_0(\operatorname{Im} {}^t\sigma). $$

It suffices to combine Theorem 14.1 with Proposition 17.3.

Corollary 1. – *Same hypotheses as in Theorem* 17.2. *Suppose, in addition, that F is complete. Then the following properties are equivalent*:

(a) $\tilde{u}_0$ *is an epimorphism and has the homogeneous approximation property*;

(b) (b_1) *for all* $x' \in E'$,

$$ {}^t\varrho(x') \in (\operatorname{Ker} u_0)^\perp \quad \textit{implies} \quad {}^t\varrho(x') \in {}^t u_0(\operatorname{Im} {}^t\sigma); $$

(b_2) ${}^t u_0 \circ {}^t\sigma$ *is injective*;

(b_3) $\tilde{E}_0/\operatorname{Ker} \tilde{u}_0$ *is complete.*

Combine Corollary 1 of Theorem 14.1 with Proposition 17.3.

Corollary 2. – *Same hypotheses as in Theorem* 17.2. *Suppose, in addition, that the following two conditions are satisfied:*

$$ \operatorname{Im} {}^t u_0 = (\operatorname{Ker} u_0)^\perp, \tag{17.5} $$

$$ {}^t u_0 \textit{ is injective } (\text{i.e., } \operatorname{Im} u_0 \textit{ is dense in } F_0). \tag{17.6} $$

Then, the equivalent properties (a) *and* (b) *of Theorem* 17.2 *are equivalent with the following one*:

(c) *for every* $y_0' \in F_0'$,

$$ {}^t u_0(y_0') \in \operatorname{Im} {}^t\varrho \quad \textit{implies} \quad y_0' \in \operatorname{Im} {}^t\sigma. $$

Proof: We prove the equivalence of (b) in Theorem 17.2 with (c):

1°) Suppose that (b) holds, and let $y_0' \in F_0'$ be such that ${}^t u_0(y_0') = {}^t\varrho(x')$ for some $x' \in E'$. We must then have ${}^t u_0(y_0') = {}^t u_0({}^t\sigma(y'))$ for some $y' \in F'$. In view of (17.6), this implies $y_0' = {}^t\sigma(y')$, whence (c).

2°) Suppose that (c) holds and let $x' \in E'$ be such that ${}^t\varrho(x') \in (\operatorname{Ker} u_0)^\perp$. In view of (17.5), there is $y_0' \in F_0'$ such that ${}^t\varrho(x') = {}^t u_0(y_0')$ whence, by (c), ${}^t\varrho(x') \in {}^t u_0(\operatorname{Im} {}^t\sigma)$. Q.E.D.

The case where all the intervening spaces are Fréchet spaces is particularly simple. Indeed, in this case, Conditions (17.3) and (17.4) are trivially satisfied. Note, on the other hand, that $\tilde{E}_0$, equipped with the graph topology, is then also a Fréchet space and therefore, this is also true of $\tilde{E}_0/\operatorname{Ker} \tilde{u}_0$. Note also, still in the case of Fréchet spaces, that the conjunction of (17.5) & (17.6) is equivalent with the fact that u_0 is surjective (Theorem 17.1). If this is true, Property (b_2) in Corollary 1 simply means that $\operatorname{Im}\sigma$ is dense in F. Finally we may state the following important corollary of Theorem 17.2:

Corollary 3. – *Let E_0, F_0, E, F be Fréchet spaces and suppose that u_0 is essentially univalent with respect to the pair (ϱ, σ). Suppose furthermore that u_0 is surjective. Then the following properties are equivalent:*

(a) *$\tilde{u}_0$ is an epimorphism and has the homogeneous approximation property;*

(b) *$\sigma(F_0)$ is dense in F and for every $y_0' \in F_0'$,*

$$^t u_0(y_0') \in \operatorname{Im}{}^t\varrho \quad \textit{implies} \quad y_0' \in \operatorname{Im}{}^t\sigma .$$

Part II.
Applications to Linear Partial Differential Equations

Chapter VI. Applications of the Epimorphism Theorem

18. A classical theorem of E. Borel

We denote by $\mathbf{R}^n$ the n-dimensional Euclidean space, by $x=(x_1,\ldots,x_n)$ the variable point in $\mathbf{R}^n$, by $\mathbf{N}^n$ the subset of $\mathbf{R}^n$ consisting of the *n-tuples* $p=(p_1,\ldots,p_n)$, where the p_j are integers $\geqslant 0$. We set $|p|=p_1+\cdots+p_n$ (whereas $|x|=(x_1^2+\cdots+x_n^2)^{1/2}$). If f is a $\mathscr{C}^\infty$ functions in $\mathbf{R}^n$, say with complex values, we write

$$f^{(p)} \quad \text{for} \quad (\partial/\partial x_1)^{p_1}\ldots(\partial/\partial x_n)^{p_n} f\,.$$

We shall, as a matter of fact, deal with the space $\mathscr{C}^\infty(\mathbf{R}^n)$ of complex $\mathscr{C}^\infty$ functions in $\mathbf{R}^n$; this space carries a natural Fréchet space structure, namely the one defined by the seminorms

$$f \rightsquigarrow |f|_{m,K} = \sup_{x\in K} \sum_{|p|\leqslant m} |f^{(p)}(x)|\,,$$

where K is an arbitrary compact subset of $\mathbf{R}^n$, m an integer $\geqslant 0$.

We shall deal with a second Fréchet space: the space Σ of complex sequences $(s_p)_{p\in\mathbf{N}^n}$, taking as a basis of continuous seminorms on Σ, the seminorms

$$s=(s_p)_{p\in\mathbf{N}^n} \rightsquigarrow |s|_m = \sum_{|p|\leqslant m} |s_p| \qquad (m=0,1,\ldots)\,.$$

The dual of the Fréchet space Σ can be identified with the space Σ^F of the finite sequences $s'=(s_p)_{p\in\mathbf{N}^n}$ (finite means that $s'_p=0$ for $|p|$ larger than some integer), via the duality bracket

$$\langle s', s\rangle = \sum_p s'_p s_p\,. \tag{18.1}$$

It is clear that $s' \rightsquigarrow (s \rightsquigarrow \langle s', s\rangle)$ defines an injection of Σ^F into Σ'. Conversely, if L is a linear functional on Σ, we may set

$$s'_p = L(e_p)\,, \tag{18.2}$$

where e_p is the sequence having all terms equal to zero except the p-th one, equal to one. If L is continuous, we have, for some integer $m\geqslant 0$

and some constant $C>0$, and for all $s\in\Sigma$,

$$|L(s)| \leqslant C|s|_m.$$

This implies at once that s'_p, defined by (18.2), vanishes if $|p|>m$, hence the sequence $s'=(s'_p)_{p\in\mathbf{N}^n}$ is finite. Moreover,

$$L(s)=\langle s', s\rangle, \qquad s\in\Sigma.$$

We associate with the finite sequence s' the polynomial

$$P_{s'}(X)=\sum_p s'_p X^p \qquad (X^p = X_1^{p_1}\dots X_n^{p_n}).$$

Suppose that, for each p, $s_p=(i\xi)^p$ with $\xi\in\mathbf{R}^n$ $(i=\sqrt{-1})$. We have then

$$\langle s', s\rangle = P_{s'}(i\xi). \tag{18.3}$$

Observe that there is a natural mapping of $\mathscr{C}^\infty(\mathbf{R}^n)$ into Σ:

$$f \rightsquigarrow (f^{(p)}(0))_{p\in\mathbf{N}^n}. \tag{18.4}$$

This mapping is linear and continuous. Let us denote it by u. Observe that, if f is the function

$$x\rightsquigarrow \exp(i\langle x,\xi\rangle), \qquad \xi\in\mathbf{R}^n, \tag{18.5}$$

then $u(f)$ is the sequence $((i\xi)^p)_{p\in\mathbf{N}^n}$ and therefore, in view of (18.3), we have, for all $s'\in\Sigma^F$,

$$\langle s', u(f)\rangle = P_{s'}(i\xi). \tag{18.6}$$

Borel's theorem states essentially that the mapping (18.4) is surjective:

Theorem 18.1. – *Let $s=(s_p)_{p\in\mathbf{N}^n}$ be an arbitrary sequence in n indices, with complex terms. Let x^0 be an arbitrary point of $\mathbf{R}^n$. There exists a function $f\in\mathscr{C}^\infty(\mathbf{R}^n)$ such that, for all $p\in\mathbf{N}^n$,*

$$f^{(p)}(x^0)=s_p.$$

Proof: We take $x^0=0$. Let K be an arbitrary compact subset of $\mathbf{R}^n$ and m any integer $\geqslant 0$. Let s' be a finite sequence such that for some constant $C>0$ and for all $f\in\mathscr{C}^\infty(\mathbf{R}^n)$,

$$|\langle s', u(f)\rangle| = |\langle {}^t u(s'), f\rangle| \leqslant C|f|_{m,K} \tag{18.7}$$

(u denotes the mapping (18.4) and ${}^t u$, its transpose). We take for f the function (18.5) and we obtain, by combining (18.6) and (18.7),

$$|P_{s'}(i\xi)| \leqslant C \sum_{|p|\leqslant m} |\xi^p|.$$

This means that $s'_p = 0$ for $|p| > m$ and that $\sum_p |s'_p|$ is bounded by a constant depending only on $C > 0$. This implies what we wanted, as we may apply Theorem 17.1 (equivalence of (e) with (a)).

19. Estimates in Sobolev spaces leading to the existence of solutions to a linear partial differential equation

We recall rapidly the definition of the space H^s (s: real number): it is the space of tempered distributions in $\mathbf{R}^n$, u, whose Fourier transform $\hat{u}$ is a square integrable function with respect to the measure

$$(1 + |\xi|^2)^s \, d\xi$$

(we denote by ξ the variable in $\mathbf{R}^n$ on the Fourier transform side). We equip H^s with the inner product

$$(u, v)_s = \int_{\mathbf{R}^n} \hat{u}(\xi) \, \bar{\hat{v}}(\xi) (1 + |\xi|^2)^s \, d\xi,$$

and with the associated norm

$$\|u\|_s = \left(\int_{\mathbf{R}^n} |\hat{u}(\xi)|^2 (1 + |\xi|^2)^s \, d\xi \right)^{1/2}.$$

They turn it into a Hilbert space. The dual of H^s is canonically isomorphic (as a Hilbert space) with H^{-s}; the duality bracket is

$$(u, v) \rightsquigarrow \langle u, v \rangle = \int \hat{u}(\xi) \, \hat{v}(-\xi) \, d\xi$$

which extends the bilinear form $\int \varphi \psi dx$ on $\mathscr{C}_c^\infty(\mathbf{R}^n) \times \mathscr{C}_c^\infty(\mathbf{R}^n)$. When s is an integer $\geqslant 0$, H^s is the space of distributions whose derivatives of order $\leqslant s$ belong to L^2. When s is an integer < 0, it is the space of distributions which are sums of derivatives of order $\leqslant |s|$ of L^2 functions. For more information on the subject of the spaces H^s, see e.g. TREVES [*1*], Section 0.6.

Let Ω be an open subset of $\mathbf{R}^n$, $\mathscr{C}^\infty(\Omega)$ the space of $\mathscr{C}^\infty$ functions (with complex values) in Ω. Consider a sequence of $\mathscr{C}^\infty$ function g_j ($j = 0, 1, \ldots$) with compact support in Ω, such that the interior Ω^j of the set $\{x; g_j(x) = 1\}$ contains supp g_{j-1} for each $j \geqslant 1$, and such that the Ω^j (which form obviously an increasing sequence of relatively compact open subsets of Ω) fill up Ω. Then the usual $\mathscr{C}^\infty$ topology on $\mathscr{C}^\infty(\Omega)$ may be defined by the seminorms

$$f \rightsquigarrow \|g_j f\|_k, \qquad j, k = 0, 1, \ldots.$$

This turns $\mathscr{C}^\infty(\Omega)$ into a Fréchet space whose dual "is" the space $\mathscr{E}'(\Omega)$ of distributions with compact support in Ω. Let μ be such a distribution, r any real number.

Lemma 19.1. – *Suppose that, for some integer $j \geqslant 0$ and some constant $C > 0$, we have:*

$$\text{for all } f \in \mathscr{C}^\infty(\Omega), \quad |\langle \mu, f\rangle| \leqslant C \|g_j f\|_r. \tag{19.1}$$

Then we have:

$$\operatorname{supp} \mu \subset K, \tag{19.2}$$

$$\mu \in H^{-r} \quad \text{and} \quad \|\mu\|_{-r} \leqslant C', \tag{19.3}$$

with $K = \operatorname{supp} g_j$ and

$$C' = C \sup_{\xi \in \mathbf{R}^n} |\hat{g}_j(\xi)|. \tag{19.4}$$

Conversely, if (19.2) *holds for some compact subset K of Ω and if* (19.3) *holds for some constant $C' > 0$, then* (19.1) *holds with $C = C'$ and $g_j = 1$ in some neighborhood of K.*

Proof: it is apparent in the statement itself. It suffices to point out that if (19.1) holds, and if f is the restriction to Ω of a function belonging to $\mathscr{C}_c^\infty(\mathbf{R}^n)$, say $\tilde{f}$, we have

$$|\langle \mu, \tilde{f}\rangle| = |\langle \mu, f\rangle| \leqslant C \|g_j f\|_r \leqslant C' \|\tilde{f}\|_r,$$

where C' is given by (19.4). The density of $\mathscr{C}_c^\infty(\mathbf{R}^n)$ in H^r implies at once (19.3).

We consider now a *differential operator* in Ω, $\mathbf{P}$, that is to say, a linear partial differential operator with $\mathscr{C}^\infty$ (complex) coefficients in Ω. We regard $\mathbf{P}$ as a continuous linear map of $\mathscr{C}^\infty(\Omega)$ into itself; as such, it has a transpose

$$^t\mathbf{P}: \mathscr{E}'(\Omega) \to \mathscr{E}'(\Omega).$$

It is very easy, at this stage, to obtain the following result, due to Harvey [*1*]:

Theorem 19.1. – *The continuous linear map $\mathbf{P}: \mathscr{C}^\infty(\Omega) \to \mathscr{C}^\infty(\Omega)$ is surjective if and only if the following condition is fulfilled:*

(*) *To every compact subset K of Ω and to every real number s, there is another compact subset $K' \subset \Omega$, a real number t and a constant $B > 0$ such that, for all $\mu \in \mathscr{E}'(\Omega)$, the property*

$$^t\mathbf{P}\mu \in H^s \quad \text{and} \quad \operatorname{supp} {}^t\mathbf{P}\mu \subset K, \tag{19.5}$$

implies the properties:

$$\operatorname{supp}\mu \subset K', \tag{19.6}$$

$$\mu \in H^t \quad and \quad \|\mu\|^t \leqslant B \, \|{}^t\mathbf{P}\mu\|_s . \tag{19.7}$$

Proof: Combine Proposition 17.1 with Lemma 19.1 (as $\mathscr{C}^\infty(\Omega)$ is a Fréchet space, presurjectivity implies surjectivity, in view of Theorem 10.1).

Estimates of the kind (19.7) have been established for a variety of types of differential operators; see, for instance, HÖRMANDER [*1*], Theorem 8.7.1.

We have restricted ourselves to the case of a single linear partial differential equation $\mathbf{P}u=v$ in an open subset of the Euclidean space. There was no need, other than for simplicity sake, for such a restriction: Theorem 19.1 extends trivially to systems of linear PDE's over a $\mathscr{C}^\infty$ manifold M, provided that the latter be countable at infinity. As a matter of fact, we could have considered two vector bundles V, W, over the manifold M, and a V–W-differential operator $\mathbf{P}$. The fibres in V and W do not even have to be finite dimensional! The theorem is still valid if the fibres are infinite dimensional Hilbert or even Banach spaces. For such generalizations, one should simply be careful in dealing with the H^s spaces.

We also wish to underline the following fact. The property that, given any $\mathscr{C}^\infty$ function v, we may find a $\mathscr{C}^\infty$ function u such that $\mathbf{P}u=v$, does not mean that we may find u such that the mapping $v \rightsquigarrow u$ be a continuous linear map of $\mathscr{C}^\infty$ into itself. The latter property is expressed by saying that $\mathbf{P}$ has a continuous right inverse. It does not hold in important instances, for example when $\mathbf{P}$ is an elliptic operator with analytic coefficients (see Appendix C, Theorem C.1). It does however hold in other cases, for example when $\mathbf{P}$ is a hyperbolic operator, say (for simplicity) with constant coefficients in $\mathbf{R}^n$, and if the equation $\mathbf{P}u=v$ is considered in the whole space $\mathbf{R}^n$: then we may take for u the unique solution of the Cauchy problem which has Cauchy data equal to zero on some non-characteristic hyperplane (chosen once for ever: it should not depend on u!); with this choice of u, the mapping $v \rightsquigarrow u$ is continuous.

20. P-convexity and semiglobal solvability

We continue to deal with the same objects as in Section 19: an open subset Ω of the Euclidean space $\mathbf{R}^n$, a differential operator $\mathbf{P}$ defined in Ω.

Definition 20.1. – *We say that the open set Ω is* **P**-*convex if, to every compact subset K of Ω and to every real number s, there is another compact subset K' of Ω such that, for all distributions μ with compact support in Ω,*

$$ {}^t\mathbf{P}\mu \in H^s, \quad \operatorname{supp} {}^t\mathbf{P}\mu \subset K \quad \textit{implies} \quad \operatorname{supp} \mu \subset K'. $$

The name "**P**-convexity" has been introduced by L. HÖRMANDER but the property, in a somewhat more primitive form (adapted to situation under study), and its incidence on existence of solutions of PDE's, was discovered by B. MALGRANGE in his thesis [*1*]. Definition 20.1 provides the motivation for the definition of u-convex pairs (E, F) (Definition 11.1). Indeed:

Proposition 20.1. – *The following properties are equivalent:*
(a) *the open set Ω is* **P**-*convex*;
(b) *the pair $(\mathscr{C}^\infty(\Omega), \mathscr{C}^\infty(\Omega))$ is* **P**-*convex.*

Proof: (a)$\Rightarrow$(b). We use, as a basis of continuous seminorms on $\mathscr{C}^\infty(\Omega)$, the seminorms $f \rightsquigarrow \|g_j f\|_k$ $(j, k=0, 1, \ldots)$ defined in p. 60. Let $\mu \in \mathscr{E}'(\Omega)$ be such that, for some constant $C>0$ and all $f \in \mathscr{C}^\infty(\Omega)$,

$$ |\langle {}^t\mathbf{P}\mu, f\rangle| \leqslant C \|g_j f\|_{-s}. $$

This implies (cf. Lemma 19.1) that $\operatorname{supp} {}^t\mathbf{P}\mu \subset K = \operatorname{supp} g_j$ and that ${}^t\mathbf{P}\mu \in H^s$. In view of (a), there is a compact subset K' of Ω such that $\operatorname{supp} \mu \subset K'$. Let then U be a relatively compact open neighborhood of K' in Ω and set, for all $f \in \mathscr{C}^\infty(\Omega)$,

$$ \mathfrak{Q}(f) = \sup_{x \in U} |f(x)|. $$

We have obviously $\operatorname{Ker} \mathfrak{Q} \subset \operatorname{Ker} \mu$, whence (b) in view of Proposition 17.2.

(b)$\Rightarrow$(a). Let $\mu \in \mathscr{E}'(\Omega)$ be such that ${}^t\mathbf{P}\mu \in H^s$ and $\operatorname{supp} {}^t\mathbf{P}\mu \subset K$. Again in view of Lemma 19.1, we have, for some constant $C>0$ and for all $f \in \mathscr{C}^\infty(\Omega)$,

$$ |\langle {}^t\mathbf{P}\mu, f\rangle| \leqslant C \|g_j f\|_{-s}, $$

where $g_j = 1$ in some neighborhood of K. If (b) holds, again in virtue of Proposition 17.2, there is a continuous seminorm $\mathfrak{Q}$ on $\mathscr{C}^\infty(\Omega)$, depending only on s and j, such that $\operatorname{Ker} \mathfrak{Q} \subset \operatorname{Ker} \mathfrak{Q}$. But $\operatorname{Ker} \mathfrak{Q}$ contains the kernel of a seminorm

$$ f \rightsquigarrow \|g_{j'} f\|_k $$

for some j' and k, hence we must have $\operatorname{supp} \mu \subset \operatorname{supp} g_{j'}$. Q.E.D.

We have already seen (Theorem 19.1) that the surjectivity of $\mathbf{P}: \mathscr{C}^\infty(\Omega) \to \mathscr{C}^\infty(\Omega)$ implies that Ω is **P**-convex.

Definition 20.2. – *We say that the partial differential equation*

$$\mathbf{P}u = v \tag{20.1}$$

is semiglobally solvable in $\mathscr{C}^\infty(\Omega)$ *if, for every relatively compact open subset* Ω' *of* Ω, *and every function* $v \in \mathscr{C}^\infty(\Omega)$, Eq. (20.1) *is satisfied, in* Ω', *by some* $u \in \mathscr{C}^\infty(\Omega)$.

Proposition 20.2. – *The following properties are equivalent:*

(a) *Eq.* (20.1) *is semiglobally solvable in* $\mathscr{C}^\infty(\Omega)$;

(b) *the continuous linear map*

$$\mathbf{P}: \mathscr{C}^\infty(\Omega) \to \mathscr{C}^\infty(\Omega)$$

is semiglobally surjective (Definition 11.2).

Proof: It suffices to make the following observation: to every continuous seminorm $\mathfrak{Q}$ on $\mathscr{C}^\infty(\Omega)$, there is a relatively compact open subset Ω' of Ω such that

$$f = 0 \text{ in } \Omega' \quad \text{implies} \quad f \in \operatorname{Ker} \mathfrak{Q},$$

and conversely, for every relatively compact open subset Ω' of Ω, there is a continuous seminorm $\mathfrak{Q}$ on $\mathscr{C}^\infty(\Omega)$ whose kernel consists exactly of the functions vanishing in Ω' (for example, $\mathfrak{Q}(f) = \sup_{x \in \Omega'} |f(x)|$).

The choice of the adjective "semiglobal" is motivated by the fact that the solvability of Eq. (20.1) under consideration is less than global, evidently, but more than local, which would mean that each point of Ω has a neighborhood where (20.1) is solvable. We shall see, later on, that the latter property is really not equivalent with semiglobal solvability.

Taking into account Propositions 20.1 & 20.2, we may apply Corollary 1 of Proposition 11.2:

Theorem 20.1. – *The following properties are equivalent:*

(a) $\mathbf{P}\mathscr{C}^\infty(\Omega) = \mathscr{C}^\infty(\Omega)$;

(b) Ω *is* **P**-*convex and the equation* $\mathbf{P}u = v$ *is semiglobally solvable in* $\mathscr{C}^\infty(\Omega)$.

Let us introduce the following definition:

Definition 20.3. – *Let* K *be a compact subset of the open set* Ω, s *any real number. We shall call* **P**-*convex hull of degree* s *of* K *in* Ω *the union of the sets* $\operatorname{supp} \mu$, *when* μ *runs over the set of distributions with compact support in* Ω, *such that* ${}^t\mathbf{P}\mu \in H^s$ *and* $\operatorname{supp} {}^t\mathbf{P}\mu \subset K$.

We shall denote by $\Gamma^s(K; \mathbf{P}, \Omega)$ the **P**-convex hull of degree s of K in Ω.

Proposition 20.3. – *Suppose that* $\mathbf{P}\mathscr{C}^\infty(\Omega) = \mathscr{C}^\infty(\Omega)$. *For every compact set* $K \subset \Omega$ *and every real number* s, *the* **P**-*convex hull of degree* s *of* K *in* Ω *is compact. If* s *is sufficiently close to* $-\infty$ *it contains* K.

Proof: We assume throughout the proof that $\mathbf{P}\mathscr{C}^\infty(\Omega)=\mathscr{C}^\infty(\Omega)$. Then, in view of Theorem 20.1, we know that $\Gamma^s=\Gamma^s(K;\mathbf{P},\Omega)$ is relatively compact in Ω; we must therefore show that Γ^s is closed. We shall reason by contradiction and assume that there is a point $x\in\Omega$, belonging to the closure of Γ^s but not to Γ^s. This means that there is a sequence of open neighborhoods $\{U_j\}$ of x and a sequence of distributions $\mu_j\in\mathscr{E}'(\Omega)$ $(j=1,2,\ldots)$ such as to have, for every j, $\overline{U_{j+1}}\subset U_j$ and

$$ {}^t\mathbf{P}\mu_j\in H^s(K); \tag{20.2}$$

$$ (\operatorname{supp}\mu_j)\cap U_j\neq\emptyset \quad\text{and}\quad (\operatorname{supp}\mu_j)\cap U_{j+1}=\emptyset. \tag{20.3}$$

We may, and we shall moreover assume that, for all j,

$$\|{}^t\mathbf{P}\mu_j\|_s\leqslant 1. \tag{20.4}$$

For each j, we select a function $\varphi_j\in\mathscr{C}_c^\infty(\Omega)$ such that

$$\operatorname{supp}\varphi_j\subset U_j; \tag{20.5}$$

$$\langle\mu_j,\varphi_j\rangle=1. \tag{20.6}$$

From (20.5) follows that $\langle\mu_i,\varphi_j\rangle=0$ for all $i<j$.

We construct then, by induction on $j=1,2,\ldots$, a sequence of numbers $c_j>0$, in the following manner:

$$c_1=1;\quad\text{for}\quad j>1,\quad c_j=2^{-j}\inf_{1\leqslant i\leqslant j-1}\frac{c_i}{1+|\langle\mu_j,\varphi_i\rangle|}.$$

From the fact that $c_j\leqslant 2^{-j}$ for all j, and from (20.4), it follows at once that the series

$$\sum_{j=1}^{+\infty} c_j{}^t\mathbf{P}\mu_j$$

converges absolutely in the Banach space $H^s(K)$. Let v be its sum. We apply Theorem 19.1: we obtain that the series

$$\sum_{j=1}^{+\infty} c_j\mu_j$$

converges absolutely in $H^t(K')$ for a suitable real number t and a suitable compact subset K' of Ω; let μ be its sum. We have ${}^t\mathbf{P}\mu=v\in H^s(K)$. What is left is to show that $x\in\operatorname{supp}\mu$: we will have then reached a contradiction.

For all $j=1,2,\ldots$, we have

$$\langle\mu,\varphi_j\rangle=c_j+\sum_{k>j}c_k\langle\mu_k,\varphi_j\rangle.$$

From our definition of the numbers c_j we derive at once that $|\langle\mu,\varphi_j\rangle|\geqslant c_j/2$ for all j, hence $x\in\operatorname{supp}\mu$.

As for the fact that $K \subset \Gamma^s(K; \mathbf{P}, \Omega)$ if s is sufficiently close to $-\infty$, it suffices to observe that, for every $x \in K$, ${}^t\mathbf{P}\delta_x \in H^s$ as soon as $s < -\omega - n/2$, where δ_x denotes the Dirac measure at the point x and ω the order of ${}^t\mathbf{P}$ in some relatively compact neighborhood of K (TREVES [1], Corollary of Theorem 0.44). We have $\operatorname{supp} {}^t\mathbf{P}\delta_x \subset \{x\}$. Q.E.D.

We shall make further remarks on **P**-convexity and semiglobal solvability in the next section. Let us underline that here also, as in Section 19, there was no need to restrict ourselves to an open subset of $\mathbf{R}^n$ and to a scalar differential operator: all the definitions and the results of the present section extend to fibre bundles over a manifold M, V and W, and to V–W-differential operators.

21. Remarks on P-convexity and semiglobal solvability

There are cases where the application of Theorem 20.1, i.e., the verification of semiglobal solvability, is considerably easier than the application of Theorem 19.1, i.e., the proof of a sequence of estimates in the Sobolev spaces H^s. A conspicuous example of such cases is provided by the linear PDE with constant coefficients. Next chapter is devoted to its study. But there are other situations, of which we study now an example:

21.1. *The case of hypoelliptic operators*

As before, we consider the open subset Ω of the Euclidean space $\mathbf{R}^n$, and the differential operator $\mathbf{P}$ in Ω.

Definition 21.1. – *Let T be a distribution in Ω. The singular support of T is the smallest closed subset of Ω, in whose complement T is a $\mathscr{C}^\infty$ function.*

We shall denote by

$$\operatorname{sing\,supp} T$$

the singular support of T.

Definition 21.2. – *The differential operator* $\mathbf{P}$ *in Ω is said to be hypoelliptic if, for every distribution T in Ω,*

$$\operatorname{sing\,supp} \mathbf{P}T = \operatorname{sing\,supp} T . \tag{21.1}$$

Observe that we have always

$$\operatorname{sing\,supp} T \supset \operatorname{sing\,supp} \mathbf{P}T .$$

Another way of expressing (21.1) is by saying that T is a $\mathscr{C}^\infty$ function in every open subset of Ω where this is true of $\mathbf{P}T$.

Examples of hypoelliptic operators: all the elliptic operators and the parabolic ones.

The hypoelliptic operators with constant coefficients have been characterized by HÖRMANDER (see HÖRMANDER [*1*], Ch. IV, also: TREVES [*1*], Ch. 7).

The property of the hypoelliptic operators, of interest to us in the present context, is the following one:

Proposition 21.1. – *Let* $\mathbf{P}$ *be a hypoelliptic differential operator in* Ω. *Suppose that, to every relatively compact open subset* Ω' *of* Ω *there are real numbers* s, t *and a constant* $C>0$ *such that*

$$\|\varphi\|_s \leqslant C\,\|{}^t\mathbf{P}\varphi\|_t \quad \text{for all} \quad \varphi\in\mathscr{C}_c^\infty(\Omega'). \tag{21.2}$$

Then the equation

$$\mathbf{P}u = v \tag{21.3}$$

is semiglobally solvable in $\mathscr{C}^\infty(\Omega)$ (Definition 20.1).

Proof: It suffices to prove the following: given an arbitrary relatively compact open subset Ω' of Ω, to every $\varphi\in\mathscr{C}^\infty(\Omega')$ there is $\mu\in\mathscr{D}'(\Omega')$ such that $\mathbf{P}\mu=\varphi$. The hypoellipticity of $\mathbf{P}$ implies at once that μ is then a $\mathscr{C}^\infty$ function in Ω'. Now, if v is an arbitrary $\mathscr{C}^\infty$ function in Ω, and if Ω'' is an arbitrary relatively compact open subset of Ω, we take Ω' to be an open (and relatively compact) neighborhood of $\overline{\Omega''}$ in Ω, and $\varphi=gv$, where $g\in\mathscr{C}_c^\infty(\Omega')$ is equal to one in a neighborhood of $\overline{\Omega'}$; then $u=g\mu\in\mathscr{C}_c^\infty(\Omega')$ satisfies (21.3) in Ω''. Thus Proposition 21.1 will be a consequence of the following result:

$$\textit{If (21.2) holds, to every } \varphi\in\mathscr{C}_c^\infty(\Omega') \textit{ there is } \mu\in\mathscr{D}'(\Omega') \textit{ such that } \mathbf{P}\mu=\varphi \textit{ in } \Omega'. \tag{21.4}$$

Assertion (21.4) is always true – even when the differential operator $\mathbf{P}$ is not hypoelliptic. It is a straightforward application of an elementary lemma about normal Hilbert spaces of distributions, stated and proved in Appendix B (Lemma B.1). In the present case, the Hilbert spaces under consideration are the following ones: $\mathscr{H}_1$ (resp., $\mathscr{H}_2$) is the closure in H^s (resp. H^t) of $\mathscr{C}_c^\infty(\Omega')$ (we deal here with spaces of distributions in Ω'). Note that $\mathscr{C}_c^\infty(\Omega')\subset(H^s)'$ (cf. (B.1)). Q.E.D.

Proposition 21.1 shows that, when the differential operator under study, $\mathbf{P}$, is hypoelliptic, there is no need to prove a whole string of estimates for each $\Omega'\subset\subset\Omega$, as requested in Theorem 19.1: it suffices to prove a single estimate for each Ω' and this estimate is, in fact, considerably simpler than the estimates (19.7), as it bears only on test functions φ and not on arbitrary distributions belonging to H^s with support in Ω' (and whose image under ${}^t\mathbf{P}$ belongs to H^t). A glance at the methods of proving estimates, known at the present time (e.g., see HÖRMANDER [1], proof of Theorem 8.7.1), will show that this is not a minor simplification. Let us recall that Proposition 21.1 applies in particular to elliptic and parabolic operators.

21.2. *Local solvability versus semiglobal solvability*

Let Ω be an open subset of $\mathbf{R}^n$, $\mathbf{P}$ a differential operator in Ω.

Definition 21.3. – *We say that the equation*

$$\mathbf{P}u = v \tag{21.5}$$

is locally solvable in $\mathscr{C}^\infty(\Omega)$ if every point $x^0 \in \Omega$ has an open neighborhood $U(x^0) \subset \Omega$ such that, for every $v \in \mathscr{C}^\infty(\Omega)$, Eq. (21.5) is satisfied in $U(x^0)$ by some $u \in \mathscr{C}^\infty(\Omega)$.

There are linear PDE's which are not locally solvable (cf. Proposition 21.4). Here we shall give a very simple example of a linear PDE which is locally, but not semiglobally, solvable.

Proposition 21.2. – *Let Ω be the complement of the origin in the plane $\mathbf{R}^2$. The equation $(\partial/\partial\theta)u = v$, where*

$$\partial/\partial\theta = (x_1^2 + x_2^2)^{-1}[x_1(\partial/\partial x_2) - x_2(\partial/\partial x_1)]$$

is locally, but not semiglobally, solvable in $\mathscr{C}^\infty(\Omega)$.

Proof: Switching to polar coordinates (r,θ) in $\Omega = \mathbf{R}^2 \setminus \{0\}$ shows at once that $\partial/\partial\theta$ is locally solvable in $\mathscr{C}^\infty(\Omega)$. The transpose of $\partial/\partial\theta$ is the operator $-\partial/\partial\theta : \mathscr{E}'(\Omega) \to \mathscr{E}'(\Omega)$, which is not injective, as $(\partial/\partial\theta)\mu = 0$ for all distributions μ which are rotation invariant: now, there are such distributions having compact support contained in Ω (e.g., the characteristic function of any anulus $0 < r_1 < r < r_2 < +\infty$). Therefore the image of $\partial/\partial\theta : \mathscr{C}^\infty(\Omega) \to \mathscr{C}^\infty(\Omega)$ is not dense, which would be the case if this operator were semiglobally surjective (Definition 11.2).

21.3. *Independence of* P*-convexity and semiglobal solvability*

Let Ω and $\mathbf{P}$ be like before, in §21.2. We begin by observing that a linear PDE *with constant coefficients* in $\mathbf{R}^n$ is always semiglobally solvable in $\mathscr{C}^\infty(\mathbf{R}^n)$. It suffices to use a fundamental solution of the equation (see proof of Theorem 23.1). In view of this, the next result exhibits a situation where semiglobal solvability holds but not $\mathbf{P}$-convexity:

Proposition 21.3. – *The complement of the origin in the plane* $\mathbf{R}^2$ *is not* $(\partial/\partial x_1)$*-convex.*

Proof: For $\varepsilon>0$, let χ_ε be the characteristic function of the set

$$\{(x_1, x_2)\in\mathbf{R}^2;\ x_1^2+x_2^2\leqslant 1,\ x_1\geqslant\varepsilon\}.$$

Then $\operatorname{supp}(\partial/\partial x_1)\chi_\varepsilon$ is contained in the unit circonference $x_1^2+x_2^2=1$, hence in a fixed compact subset of $\mathbf{R}^2\smallsetminus\{0\}$, but this is not true of $\operatorname{supp}\chi_\varepsilon$. Now, $-\partial/\partial x_1$ is the transpose of $\partial/\partial x_1$.

The *Lewy operator*

$$L=(\partial/\partial x_1)+i(\partial/\partial x_2)+i(x_1+ix_2)(\partial/\partial x_3)$$

provides an example of validity of **P**-convexity without even local solvability. As a matter of fact, whatever (hence however small) the open set $\mathcal{O}\subset\mathbf{R}^3$, we do *not* even have $L\mathscr{D}'(\mathcal{O})\supset\mathscr{C}_c^\infty(\mathcal{O})$ (HÖRMANDER [*1*], p. 164).

Proposition 21.4. – *Let L be the Lewy operator. The open set*

$$\Omega=\{(x_1,x_2,x_3)\in\mathbf{R}^3;\ x_1>0\}\tag{21.6}$$

is L-convex.

Proof: We apply the following generalization of Holmgren's theorem due to HÖRMANDER [*1*], Lemma 5.3.2:

Lemma 21.1. – *Let* Ω *be an open subset of* $\mathbf{R}^n$, $\mathbf{P}$ *a differential operator with analytic coefficients in* Ω, *of order m, such that the coefficients of* $(\partial/\partial x_j)^m$ *never vanishes in* Ω.

For t real, let Ω_t *denote the set* $\{x\in\Omega;\ x_j<t\}$. *Then, for all distributions* μ *in* Ω,

$$\mathbf{P}\mu=0 \text{ in } \Omega_t \quad\text{and}\quad (\operatorname{supp}\mu)\cap\Omega_t\subset\subset\Omega$$

implies $\mu=0$ *in* Ω_t.

As it is customary, $A\subset\subset\Omega$ means that the closure of A is compact and contained in Ω.

We apply Lemma 21.1 with $\mathbf{P}={}^tL$, and we derive from it that, for all $\mu\in\mathscr{E}'(\Omega)$ (Ω is given by (21.6)), $\operatorname{supp}\mu$ is contained in the smallest "parallelepiped" having its faces parallel to the coordinates planes and containing $\operatorname{supp}{}^tL\mu$, whence Proposition 21.4.

21.4. *Relation with uniqueness in the Cauchy problem*

The so-called *uniqueness in the Cauchy problem* might yield some information about the **P**-convexity of an open set Ω. Let us first recall its meaning. We continue to deal with the differential operator **P** with coefficients defined and $\mathscr{C}^\infty$ in Ω. Let x^0 be some point of Ω, U an open neighborhood (in Ω) of x^0 and F a relatively closed subset of U whose boundary contains x^0. We may say that there is uniqueness in the Cauchy problem for the operator $\mathbf{P}$, at the boundary of F at the point x^0, if there is an open neighborhood $U' \subset U$ of x^0 such that the following holds:

(UCP) *For every* $\mathscr{C}^\infty$ *function* f *in* U *such that* $\mathbf{P}f=0$ *in* U *and* $f=0$ *in* $U \setminus F$, *we have* $f=0$ *in* U'.

There are several variants of this definition, also called uniqueness in the Cauchy problem. For instance, we may want to consider more general solutions of the equation $\mathbf{P}f=0$ than just the infinitely differentiable ones: if m is the order of $\mathbf{P}$ in U, we may want to consider the $\mathscr{C}^m$ solutions – or the solutions whose derivatives of order $\leqslant m$ (in distribution sense) belong to $L^2(\Omega)$. The latter is a good class, as it is particularly adapted to the exploitation of a priori estimates in the space L^2. In some favorable circumstances, we may even have the property (UCP) for all *distributions* solutions of $\mathbf{P}f=0$. An important example of such a situation has been encountered in Lemma 21.1. As a matter of fact, the use that has been made of Lemma 21.1 in the proof of Proposition 21.4 is typical of how uniqueness in the Cauchy problem has implications concerning $\mathbf{P}$-convexity. The way this works in general goes more or less as follows:

In practice, one tries always to show that, given an arbitrary compact subset K of Ω, there is another subset $\hat{K}$ of Ω such that, for all distributions $\mu \in \mathscr{E}'(\Omega)$,

$$\operatorname{supp} {}^t\mathbf{P}\mu \subset K \quad \textit{implies} \quad \operatorname{supp} \mu \subset \hat{K}.$$

One deals usually with an operator $\mathbf{P}$ which is defined in some neighborhood of $\bar{\Omega}$. At any event, at the start, one must have at least one piece of information: namely that there is a *bounded* open subset Ω' of Ω such that

$$\operatorname{supp} {}^t\mathbf{P}\mu \subset K \quad \textit{implies} \quad \operatorname{supp} \mu \subset \Omega'.$$

For instance, when the coefficients of $\mathbf{P}$ are constant, we may take, as set Ω', the intersection of Ω with any open bounded neighborhood of the convex hull of K. Here "bounded" means that Ω' is bounded in the whole space $\mathbf{R}^n$ (when reasoning on manifolds, the reasoning must be suitably modified). Naturally, we assume that $K \subset \Omega'$. Take then, as neighborhood U an open set $\Omega'' \setminus K$, where Ω'' is some open neighborhood of $\overline{\Omega'}$ in which $\mathbf{P}$ and ${}^t\mathbf{P}$ are defined, and as closed set $F \subset U$ the intersection of $\overline{\Omega'}$ with U. Suppose then that every point x^0 of the boundary $\partial\Omega'$ of Ω' has an open neighborhood $U(x^0)$ (in U) such that, if a distribution u in U satisfies ${}^t\mathbf{P}u=0$ in U, and if $u=0$ in $U \setminus F$, we have $u=0$ in $U(x^0)$. Then, for such a distribution u will vanish in the union U' of the sets $U(x^0)$ as x^0 varies over $\partial\Omega'$. Let us applies this with u, the restriction to U of a distribution $\mu\in\mathscr{E}'(\Omega')$. We see that if ${}^t\mathbf{P}\mu=0$ in U, i.e., if ${}^t\mathbf{P}\mu=0$ in $\Omega'\setminus K$, i.e., $\operatorname{supp}{}^t\mathbf{P}\mu\subset K$, then $\mu=0$ in U', i.e. $\operatorname{supp}\mu\subset\Omega'\setminus(\Omega'\cap U')$. But the complement of $\Omega'\cap U'$ with respect to Ω' is a compact subset of Ω, which we can take as set $\hat{K}$.

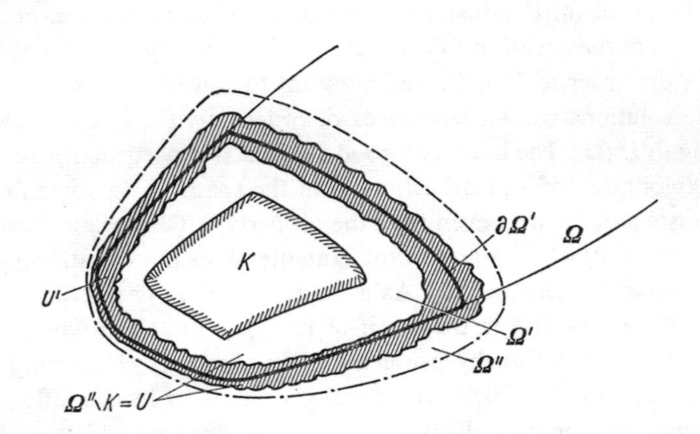

We have used the uniqueness in the Cauchy problem at the boundary of $F=U\setminus\overline{\Omega'}$ for distributions solutions of the homogeneous equation. In certain important particular cases, it will suffice to deal with $\mathscr{C}^\infty$ solutions. This is so when the operator ${}^t\mathbf{P}$ is hypoelliptic (e.g., when $\mathbf{P}$ is elliptic or parabolic). For then any distribution μ such that $\operatorname{supp}{}^t\mathbf{P}\mu\subset K$ is a $\mathscr{C}^\infty$ function in the complement of K, in particular in the above set $U=\Omega''\setminus K$.

Another case where it suffices to consider $\mathscr{C}^\infty$ solutions f of the equation ${}^t\mathbf{P}f=0$ is the case where we have *unique continuation of singularities*, for the operator ${}^t\mathbf{P}$, at the boundary of F. This means that there is an open neighborhood U' of this boundary, in U, such that, if a distribution u in U is a $\mathscr{C}^\infty$ function in $U\setminus F$, and if furthermore ${}^t\mathbf{P}u\in\mathscr{C}^\infty(U)$, then u must itself be a $\mathscr{C}^\infty$ function in U'.

A remarkable circumstance, in these questions of uniqueness in the Cauchy problem and in unique continuation of singularities, is that some positive answers can be derived from a priori estimates of the L^2 type, and that, often, the same estimates will yield semiglobal solvability (by means of Lemma B.1). On this subject, see HÖRMANDER [*1*], Theorems 8.8.1 and 8.9.1. See also TREVES [*1*], Section 2.2.

A very different approach to uniqueness in the Cauchy problem is the one that goes through the theorem of Holmgren and its variants, of which Lemma 21.1 is an example. On this subject, see HÖRMANDER [*1*], Section 5.3.

Finally, we should point out that, whereas the properties like uniqueness in the Cauchy problem are local, this is not so for **P**-convexity: it is not true that if a set is **P**-convex in some neighborhood of each one of its points, it need be **P**-convex. For a counterexample, see HÖRMANDER [*1*], p. 90.

Chapter VII. Applications of the Epimorphism Theorem to Partial Differential Equations with Constant Coefficients

22. On certain Fréchet spaces of distributions

We denote by $\mathscr{E}'$ the space of distributions in $\mathbf{R}^n$ having a compact support. We recall that the space of all the distributions in $\mathbf{R}^n$, $\mathscr{D}'$, is a convolution module over $\mathscr{E}'$: that is to say, $(\mu, \nu) \to \mu * \nu$ is a bilinear mapping of $\mathscr{D}' \times \mathscr{E}'$ into $\mathscr{D}'$.

We consider a vector subspace A of $\mathscr{D}'$ having the following properties:

A is a submodule of the $\mathscr{E}'$-convolution module $\mathscr{D}'$ (i.e., *$\mu \in A$ and $\nu \in \mathscr{E}'$ implies $\mu * \nu \in A$*); (22.1)

to every locally finite open covering $\{\mathcal{O}^j\}$ of $\mathbf{R}^n$ there is a partition of unity $\{g^j\}$ in $A \cap \mathscr{C}^\infty$, subordinated to this covering, and such that, for each index j, $g^j A \subset A$. (22.2)

Saying that the partition of unity $\{g^j\}$ is subordinated to the covering $\{\mathcal{O}^j\}$ means that, for each index j, $\operatorname{supp} g^j \subset \mathcal{O}^j$. We have denoted by $g^j A$ the set of all multiplicative products $g^j \mu$, $\mu \in A$.

Example 22.1. – The spaces $\mathscr{C}^\infty$, $\mathscr{C}_c^\infty$, $\mathscr{D}'$, $\mathscr{E}'$ satisfy (22.1) and (22.2), and so does $\mathscr{D}'^F$, the space of distributions of finite order in $\mathbf{R}^n$.

Example 22.2. – Let d be any number >1. Let us denote by $\gamma^{(d)}$ the space of $\mathscr{C}^\infty$ functions f in $\mathbf{R}^n$ having the following property:

To every compact set $K \subset \mathbf{R}^n$ and to every $\varepsilon > 0$, there is a constant $B > 0$ such that, for all n-tuples p and all $x \in K$, (22.3)

$$|(\partial/\partial x)^p f(x)| \leq B \varepsilon^{|p|} (p!)^d.$$

That $\gamma^{(d)}$ verifies (22.1) & (22.2) follows from Lemma 5.7.1 of Hörmander [*1*].

Proposition 22.1. – *Any constant coefficients differential operator in $\mathbf{R}^n$, $P(\partial/\partial x)$, maps A into itself.*

Proof: If $\mu \in A$, $P(\partial/\partial x)\,\mu = \mu * P(\partial/\partial x)\delta$, δ: Dirac measure. Thus Proposition 22.1 is a consequence of (22.1).

Proposition 22.2. – *Given an arbitrary neighborhood U of 0 in $\mathbf{R}^n$, there is a nonnegative function $\varrho \in A \cap \mathscr{C}^\infty$ such that*

$$\operatorname{supp} \varrho \subset U \quad \textit{and} \quad \int_{\mathbf{R}^n} \varrho(x)\, dx = 1.$$

Proof: Form a locally finite open covering of $\mathbf{R}^n$ of which $\mathring{U}$ (the interior of U) is an element, and such that some neighborhood of 0, V, does not intersect any other element, except $\mathring{U}$. Consider then a partition of unity in $A \cap \mathscr{C}^\infty$ subordinated to this covering (by definition of partitions of unity, the functions which constitute it are nonnegative), and let g be the function, in the partition, with support in $\mathring{U}$. Then $g=1$ in V, and it suffices to take

$$\varrho(x) = \left(\int_{\mathbf{R}^n} g(y)\, dy \right)^{-1} g(x).$$

Definition 22.1. – *Let Ω be an open subset of $\mathbf{R}^n$. We denote by*

$$A_{\text{loc}}(\Omega)$$

the space of distributions μ in Ω such that, for every relatively compact open subset Ω' of Ω, there is an element μ' of A such that $\mu = \mu'$ in Ω'.

Example 22.3. – When $A = \mathscr{C}^\infty$ (resp. $\mathscr{D}'$), we have $A_{\text{loc}}(\Omega) = \mathscr{C}^\infty(\Omega)$ (resp. $\mathscr{D}'(\Omega)$).

Example 22.4. – When $A = \gamma^{(d)}$, $d > 1$ (Example 22.2), $A_{\text{loc}}(\Omega)$ is the space $\gamma^{(d)}(\Omega)$ of functions $f \in \mathscr{C}^\infty(\Omega)$ having Property (22.3) with the modification that it involves only compact sets K contained in Ω.

We denote by

$$A_c(\Omega)$$

the subspace of A consisting of the distributions having a compact support contained in Ω. When $\Omega = \mathbf{R}^n$, we write A_{loc} and A_c rather than $A_{\text{loc}}(\mathbf{R}^n)$ or $A_c(\mathbf{R}^n)$ respectively.

Proposition 22.3. – *Let Ω' be a relatively compact open subset of Ω, K a compact neighborhood of $\overline{\Omega'}$ in Ω. There is a function $g \in A \cap \mathscr{C}^\infty$ with support in K and equal to one in Ω', such, moreover, that $g A_{\text{loc}}(\Omega) \subset A_{\text{loc}}(\Omega)$.*

Follows at once from (22.2) and Definition 22.1. We see that, for all $\mu \in A_{\text{loc}}(\Omega)$, $g\mu \in A_c(\Omega)$ and $g\mu = \mu$ in Ω'.

Proposition 22.4. – *The bilinear mapping $(\mu, \nu) \rightsquigarrow \mu * \nu$ maps $A_c \times \mathscr{D}'$ and $A_{\text{loc}} \times \mathscr{E}'$ into A_{loc}.*

Proof: Let us prove the assertion relative to $A_c \times \mathscr{D}'$. Let Ω' be an arbitrary bounded open subset of $\mathbf{R}^n$, $g \in \mathscr{C}_c^\infty$ equal one in some open neighborhood of $\operatorname{supp}\mu - \overline{\Omega'}$. Then, in Ω', $\mu * v$ equals $\mu*(gv)$ and the latter belongs to A in virtue of (22.1).

At this stage, we equip $A_{\text{loc}}(\Omega)$ with a topology submitted to the following conditions:

$$A_{\text{loc}}(\Omega) \text{ is a Fréchet space of distributions in } \Omega; \tag{22.4}$$

$$\text{the topology of } A_{\text{loc}}(\Omega) \text{ is local.} \tag{22.5}$$

We must explain the meaning of these conditions. Condition (22.4) says that the topology we put on $A_{\text{loc}}(\Omega)$ turns this space into a Fréchet space, and that the natural injection of $A_{\text{loc}}(\Omega)$ into $\mathscr{D}'(\Omega)$ is continuous. Notice that this determines the topology of $A_{\text{loc}}(\Omega)$ unambiguously. Indeed, let us denote by $\tilde{A}_{\text{loc}}(\Omega)$ the same space provided with a second topology satisfying also (22.4). Then the identity mapping of $A_{\text{loc}}(\Omega)$ onto $\tilde{A}_{\text{loc}}(\Omega)$ has a closed graph, because the injection into $\mathscr{D}'(\Omega)$ of both spaces are continuous, hence the identity mapping is continuous. Interchanging A and $\tilde{A}$, we see that the identity mapping is an isomorphism.

As for (22.5), it means that to every continuous seminorm $\mathfrak{p}$ on $A_{\text{loc}}(\Omega)$, there is a compact subset K of Ω such that, for all $\mu \in A_{\text{loc}}(\Omega)$, if μ vanishes in some neighborhood of K, then $\mathfrak{p}(\mu)=0$.

Example 22.5. – The usual $\mathscr{C}^\infty$ topology on $\mathscr{C}^\infty(\Omega)$ verifies the above condition. The strong or the weak dual topology on $\mathscr{D}'(\Omega)$ satisfies (22.5) but not (22.4). Thus, $\mathscr{D}'(\Omega)$ is outside the range of the forthcoming theorems.

Example 22.6. – Conditions (22.4) & (22.5) are satisfied by the topology defined on $\gamma^{(d)}(\Omega)$ ($d>1$, Example 22.4) by the seminorms

$$f \rightsquigarrow \sup_{p \in \mathbf{N}^n} \left(\sup_{x \in K} \{(p!)^{-d} \varepsilon^{-|p|} |(\partial/\partial x)^p f(x)|\} \right).$$

Proposition 22.5. – *The subspace $A_c(\Omega)$ is dense in $A_{\text{loc}}(\Omega)$.*

Follows at once from (22.5) and Proposition 22.3.

Let K be an arbitrary compact subset of Ω. We denote by $A_c(K)$ the subspace of $A_c(\Omega)$ consisting of the distributions with support in K. Since the injection $A_{\text{loc}}(\Omega) \to \mathscr{D}'(\Omega)$ is continuous, $A_c(K)$ is a *closed* subspace of $A_{\text{loc}}(\Omega)$, hence it becomes a Fréchet space when equipped with the induced topology. We may then provide $A_c(\Omega)$ with the topology inductive limit of $A_c(K)$ as K runs over the family of all compact subsets of Ω. A convex subset $\mathscr{U}$ of $A_c(\Omega)$ is a neighborhood of zero in this space if and only if $\mathscr{U} \cap A_c(K)$ is a neighborhood of zero in $A_c(K)$ for all K. Thus $A_c(\Omega)$ has become a space LF.

We switch now to the duals of $A_c(\Omega)$ and $A_{\text{loc}}(\Omega)$. These duals will not be, in general, spaces of distributions. But they will have essentially the same properties as $\mathscr{D}'(\Omega)$ and $\mathscr{E}'(\Omega)$ respectively. We shall put on them the strong dual topologies, that is, the topology of uniform convergence (of continuous linear functionals) on every bounded subset of the "base space", $A_c(\Omega)$ and $A_{loc}(\Omega)$ respectively.

The natural injection $A_c(\Omega)\to A_{loc}(\Omega)$ and the restriction to Ω of distributions defined in $\mathbf{R}^n$, $A_{loc}\to A_{loc}(\Omega)$, are continuous and have dense images (Proposition 22.5). Therefore, their transposes

$$(A_{loc}(\Omega))' \begin{array}{l} \nearrow (A_c(\Omega))' \\ \searrow (A_{loc})' \end{array}$$

are continuous injections. Thus $(A_{loc}(\Omega))'$ is continuously embedded, on one hand, in $(A_c(\Omega))'$, on the other hand, in $(A_{loc})'$, itself continuously embedded in $(A_c)'$. If we apply this to the choice $A=\mathscr{C}^\infty$, we find the well known situation in distribution theory: $\mathscr{E}'(\Omega)$, the space of distributions with compact support in Ω, dual of $\mathscr{C}^\infty(\Omega)$, is continuously embedded in the space $\mathscr{D}'(\Omega)$ of all the distributions in Ω, and also in $\mathscr{E}'\subset\mathscr{D}'$. This description applies word for word to the general case; in particular, the elements of $(A_{loc}(\Omega))'$ may be viewed as the elements of $(A_c(\Omega))'$ or of $(A_c)'$ having compact support – contained in Ω. In order to see this, we must simply show that the notion of support makes sense in $(A_c(\Omega))'$. Indeed it does; its definition duplicates exactly what is done in distribution theory. One begins by showing that if an element T of $(A_c)'$ vanishes in each one of a collection of open subsets $\{U^j\}$ of Ω, T vanishes in their union, which we denote by U. If $\varphi\in A_c$, $\operatorname{supp}\varphi$ is covered by a finite number of open sets U^j, say $U^{j_1},\dots, U^{j_\nu}$. In virtue of (22.2), there is a partition of unity $\{g^j\}$ $(j=0, j_1,\dots, j_\nu)$ in $A\cap\mathscr{C}^\infty$, subordinated to the covering $(U^0, U^{j_1},\dots, U^{j_\nu})$ of $\mathbf{R}^n$, where $U^0=\mathbf{R}^n\setminus\operatorname{supp}\varphi$. We have $\langle T,\varphi\rangle=\sum_{\alpha=1}^{\alpha=\nu}\langle T, g^j\varphi\rangle=0$. Q.E.D. Thus T vanishes in the union of all the open subsets of Ω in which $T=0$; the complement (with respect to Ω) of this union will be, by definition, the *support* of T. We denote it by $\operatorname{supp}T$. Condition (22.5) implies then at once that $T\in(A_c(\Omega))'$ *belongs to* $(A_{loc}(\Omega))'$ *if and only if* $\operatorname{supp}T$ *is a compact subset of* Ω.

We present now some continuity properties of convolution, which will be useful when we are going to prove the existence of solutions in $A_{loc}(\Omega)$ for linear PDE with constant coefficients (in Section 23):

Proposition 22.6. – *The bilinear mapping*

$$(\mu,\nu)\rightsquigarrow\mu*\nu$$

from $A_{loc}\times\mathscr{E}'$ *or from* $A_c\times\mathscr{D}'$ *into* A_{loc} *is separately continuous.*

Proof: As the topologies of $\mathscr{D}'$ and A_{loc} are local, it suffices to prove that the mapping under consideration is separately continuous from $A_c \times \mathscr{E}'$ into A_{loc}. This follows at once from the fact that the closed graph theorem applies to linear mappings valued in a Fréchet space (here A_{loc}) and defined either in a space LF (here A_c) or in a space which is the strong dual of a reflexive Fréchet space (here $\mathscr{E}'$; see Example 12.2).

Proposition 22.6 enables us to define the convolution of an element T of $(A_c)'$ (resp., of $(A_{\text{loc}})'$) with a distribution $\mu \in \mathscr{E}'$ (resp., $\mu \in \mathscr{D}'$). We set, for arbitrary $h \in A_c$,

$$\langle T * \mu, h \rangle = \langle T, \check{\mu} * h \rangle, \tag{22.6}$$

where $\check{\mu}$ denotes the image of μ under the symmetry with respect to the origin:

$$\langle \check{\mu}, \varphi \rangle = \langle \mu, \check{\varphi} \rangle \quad \text{and} \quad \check{\varphi}(x) = \varphi(-x), \qquad \varphi \in \mathscr{C}_c^\infty .$$

Concerning (22.6), observe that if $\mu \in \mathscr{E}'$ (resp., $\mu \in \mathscr{D}'$) $h \rightsquigarrow \check{\mu} * h$ is a continuous linear map of A_c into A_c (resp., of A_c into A_{loc}); this follows at once from Proposition 22.6 and from the fact that

$$\operatorname{supp}(\check{\mu} * h) \subset \operatorname{supp} h - \operatorname{supp} \mu .$$

The next results are immediate consequences of this definition (22.6) and from Proposition 22.6:

Proposition 22.7. – *For $T \in (A_c)'$ and $\mu \in \mathscr{D}'$, assuming that one at least of the two, T or μ, has compact support,*

$$\operatorname{supp}(T * \mu) \subset \operatorname{supp} T + \operatorname{supp} \mu .$$

In the next statements, we limit ourselves to the case where both factors, T and μ, in the convolution, have compact support. As the student may easily check, the results are still valid when one of the two (but not both) is allowed to have arbitrary support.

Proposition 22.8. – *For fixed $\mu \in \mathscr{E}'$, $T \rightsquigarrow T * \mu$ is a continuous linear map of $(A_{\text{loc}})'$ into itself.*

If $\check{\mu} \in A_c$, it is a continuous linear map of $(A_{\text{loc}})'$ into $\mathscr{C}_c^\infty$.

Concerning the last part of the statement, which might seem surprisingly strong, let us observe that $T \rightsquigarrow T * \mu$ is the transpose of the continuous linear map $h \rightsquigarrow \check{\mu} * h$. When $\check{\mu} \in A_c$, the latter is a continuous linear map of $\mathscr{D}'$ into A_{loc} (Proposition 22.6); and we recall that the dual of $\mathscr{D}'$ is $\mathscr{C}_c^\infty$.

Proposition 22.9. – *For fixed $T \in (A_{\text{loc}})'$, $\mu \rightsquigarrow T * \mu$ is a continuous linear map of $\mathscr{E}'$ into $(A_{\text{loc}})'$ when both $\mathscr{E}'$ and $(A_{\text{loc}})'$ carry their weak dual topology.*

We shall also need the following density result:

Proposition 22.10. – *The space $A_c(\Omega)\cap\mathscr{C}_c^\infty(\Omega)$ is sequentially dense in both $A_c(\Omega)$ and $\mathscr{C}_c^\infty(\Omega)$.*

Proof: Proposition 22.2 tells us that there is a sequence $\{\varrho_\varepsilon\}$ $(\varepsilon=1, \frac{1}{2}, \ldots)$ of functions belonging to $A\cap\mathscr{C}^\infty$ with support in the ball of radius ε and center at 0, such moreover that $\int\varrho_\varepsilon(x)\,dx=1$. Then $\varrho_\varepsilon*\varphi\in A_c$ for all $\varphi\in\mathscr{C}_c^\infty$ (in view of (22.1)) and converges to φ, including "supportwise", as $\varepsilon\to 0$, which proves the density in $\mathscr{C}_c^\infty(\Omega)$. Suppose now that $\varphi\in A_c$. As the ϱ_ε converge to the Dirac measure in $\mathscr{E}'$, Proposition 22.6 tells us that $\varrho_\varepsilon*\varphi$ converges to φ in A_{loc}; the fact that

$$\operatorname{supp}(\varrho_\varepsilon*\varphi)\subset\operatorname{supp}\varphi+\{x\in\mathbf{R}^n;\ |x|\leqslant\varepsilon\}$$

implies then at once that $A_c(\Omega)\cap\mathscr{C}_c^\infty(\Omega)$ is dense in $A_c(\Omega)$.

Corollary 1. – *The space $A_{\text{loc}}(\Omega)$ is sequentially dense in $\mathscr{D}'(\Omega)$.*

For $A_{\text{loc}}(\Omega)$ contains a subspace, $A_c(\Omega)\cap\mathscr{C}_c^\infty(\Omega)$, which is sequentially dense in $\mathscr{C}_c^\infty(\Omega)$, itself sequentially dense in $\mathscr{D}'(\Omega)$ (the last fact is well known; let us mention that its proof is essentially the same as the one of Proposition 22.10).

Corollary 2. – *The space $\mathscr{C}_c^\infty(\Omega)$ is canonically – and continuously – embedded in $(A_{\text{loc}}(\Omega))'$.*

The injection $\mathscr{C}_c^\infty(\Omega)\to(A_{\text{loc}}(\Omega))'$ is the transpose of the injection $A_{\text{loc}}(\Omega)\to\mathscr{D}'(\Omega)$, which has a dense image, by Corollary 1.

23. Existence of solutions to a linear PDE with constant coefficients

We continue to deal with the objects introduced in Section 22. We consider a linear partial differential operator with constant coefficients, i.e., a *differential polynomial*, in $\mathbf{R}^n$:

$$P(\partial/\partial x)=\sum_{|p|\leqslant m}c_p(\partial/\partial x)^p.$$

By p we denote an n-tuple $(p_1,\ldots,p_n)$; $|p|=p_1+\cdots+p_n$, and $(\partial/\partial x)^p=(\partial/\partial x_1)^{p_1}\ldots(\partial/\partial x_n)^{p_n}$; the *coefficients* c_p are complex numbers. In view of Proposition 22.1, $\mu\rightsquigarrow P(\partial/\partial x)\mu$ is a continuous linear map of $A_{\text{loc}}(\Omega)$ into itself: indeed, it maps this Fréchet space into itself and it is continuous if we view it as taking its values in $\mathscr{D}'(\Omega)$, hence, by the closed graph theorem, when taking its values in $A_{\text{loc}}(\Omega)$. By transposition, it gives rise to a continuous linear map of $(A_{\text{loc}}(\Omega))'$ into itself, which we denote by

$P(-\partial/\partial x)$. If $T \in (A_{\text{loc}})'$ and $\mu \in A_c$,

$$[P(-\partial/\partial x)\,T] * \check{\mu} = T * \overset{\vee}{\overbrace{P(\partial/\partial x)\,\mu}} = P(-\partial/\partial x)\,(T * \check{\mu}).$$

Notice that, in the second and third terms of this double equality, $P(\partial/\partial x)$ and $P(-\partial/\partial x)$ act in the sense of distributions.

Proposition 23.1. – *The following properties of an open subset Ω of $\mathbf{R}^n$ are equivalent:*

(a) *Ω is $P(\partial/\partial x)$-convex* (Definition 20.1);

(b) *to every compact subset K of Ω there is another compact subset K' of Ω such that, for all $\varphi \in A_c(\Omega) \cap \mathscr{C}^\infty$,*

$$\operatorname{supp} P(-\partial/\partial x)\,\varphi \subset K \quad \textit{implies} \quad \operatorname{supp} \varphi \subset K';$$

(c) *to every compact set $K \subset \Omega$ there is a compact set $K' \subset \Omega$ such that, for all $T \in (A_{\text{loc}}(\Omega))'$,*

$$\operatorname{supp} P(-\partial/\partial x)\,T \subset K \quad \textit{implies} \quad \operatorname{supp} T \subset K'.$$

Proof: (a)$\Rightarrow$(b) trivially. In view of Corollary 2 of Proposition 22.10, (c) implies (b) in which $A = \mathscr{C}^\infty$. We are going to show that (b)$\Rightarrow$(c). This will also show that the property (b) where $A = \mathscr{C}^\infty$ implies (c) with $A = \mathscr{C}^\infty$, and the latter implies trivially (a).

We suppose that (b) holds. Let $\{\varrho_\varepsilon\}$ ($\varepsilon = 1, \frac{1}{2}, \ldots$) be a sequence of functions in $A_c \cap \mathscr{C}^\infty$ such that

$$\operatorname{supp} \varrho_\varepsilon \subset B_\varepsilon = \{x \in \mathbf{R}^n;\ |x| \leqslant \varepsilon\}, \quad \int \varrho_\varepsilon(x)\,dx = 1$$

(cf. proof of Proposition 22.10). Let $T \in (A_{\text{loc}}(\Omega))'$ be such that $\operatorname{supp} P(-\partial/\partial x)\,T \subset K$. By Proposition 22.7, the support of

$$P(-\partial/\partial x)\,(T * \check{\varrho}_\varepsilon * \varrho_{\varepsilon'}) = [P(-\partial/\partial x)\,T] * \varrho_\varepsilon * \varrho_{\varepsilon'}$$

is contained in $K + B_\varepsilon + B_{\varepsilon'}$. If ε and ε' are sufficiently small, this support is contained in a fixed compact set $K_1 \subset \Omega$. Then, by Proposition 22.8, $T * \check{\varrho}_\varepsilon$ belongs to $\mathscr{C}_c^\infty$ and, for ε sufficiently small, to $\mathscr{C}_c^\infty(\Omega)$. As a consequence, if ε' is also sufficiently small, $T * \check{\varrho}_\varepsilon * \varrho_{\varepsilon'}$ belongs to $A_c(\Omega) \cap \mathscr{C}^\infty$. We may take (b) into account: there is a compact set K', depending on K_1, which contains $\operatorname{supp}(T * \check{\varrho}_\varepsilon * \varrho_{\varepsilon'})$, as soon as ε and ε' are small enough. We go to the limit $\varepsilon' = 0$; we conclude that $\operatorname{supp}(T * \check{\varrho}_\varepsilon) \subset K'$. When $\varepsilon \to 0$, the $\check{\varrho}_\varepsilon$ converge in $\mathscr{E}'$ to the Dirac measure, hence, by Proposition 22.9, $T * \check{\varrho}_\varepsilon$ converges weakly to T in $(A_{\text{loc}})'$. This implies at once that $\operatorname{supp} T \subset K'$.

Q.E.D.

Proposition 23.2. – *The equivalent conditions* (a), (b), (c) *of Proposition* 23.1 *are equivalent with the fact that the pair* $(A_{\text{loc}}(\Omega), A_{\text{loc}}(\Omega))$ *is* $P(\partial/\partial x)$-*convex* (Definition 11.1).

Proof: Condition (c) in Proposition 23.1 implies that the pair $(A_{\text{loc}}(\Omega), A_{\text{loc}}(\Omega))$ is $P(\partial/\partial x)$-convex. Indeed, let $\mathfrak{p}$ be an arbitrary continuous seminorm on $A_{\text{loc}}(\Omega)$. If $T \in (A_{\text{loc}}(\Omega))'$ satisfies, for all $\varphi \in A_{\text{loc}}(\Omega)$,

$$|\langle T, P(\partial/\partial x)\, \varphi \rangle| \leqslant \mathfrak{p}(\varphi),$$

we derive from (22.5) that $\operatorname{supp} P(-\partial/\partial x)\, T$ is contained in some compact subset K of Ω, depending only on $P(\partial/\partial x)$ and on Ω. In virtue of (c), Proposition 23.1, we derive that there is a compact subset K' of Ω, depending only on K, containing $\operatorname{supp} T$. In order to derive the desired conclusion, it will suffice to prove that there is a continuous seminorm $\mathfrak{Q}$ on $A_{\text{loc}}(\Omega)$ such that, for all $\varphi \in A_{\text{loc}}(\Omega)$, $\mathfrak{Q}(\varphi)=0$ implies that $\varphi=0$ in some neighborhood of K'. For this would imply that $T=0$ on $\operatorname{Ker} \mathfrak{Q}$, and enable us to apply Proposition 17.2. In view of this, we shall prove the following lemma:

Lemma 23.1. – *Given any compact subset* K' *of* Ω, *there is a continuous seminorm on* $\mathscr{D}'(\Omega)$, $\mathfrak{Q}$, *such that, for all* $\mu \in \mathscr{D}'(\Omega)$,

$$\mathfrak{Q}(\mu) = 0 \quad \textit{implies} \quad \operatorname{supp} \mu \subset \Omega \setminus K.$$

Proof: Let d be a number >1, Ω' a relatively compact open neighborhood of K' in Ω, $\mathfrak{B}^{(d)}(\Omega')$ the set of functions $\varphi \in \mathscr{C}_c^\infty(\Omega')$ such that, for all n-tuples p,

$$\sup_{x \in \Omega} |(\partial/\partial x)^p \varphi(x)| \leqslant (p!)^d.$$

We note that the set $\mathfrak{B}^{(d)}(\Omega')$ is bounded in $\mathscr{C}_c^\infty(\Omega)$ and, on the other hand, that it spans a vector subspace containing $\gamma_c^{(d)}(\Omega')$ (Example 22.4), which is dense in $\mathscr{C}_c^\infty(\Omega')$, in virtue of Proposition 22.10. Therefore, if a distribution in Ω vanishes on $\mathfrak{B}^{(d)}(\Omega')$, it must vanish in Ω'. Thus, the continuous seminorm on $\mathscr{D}'(\Omega)$,

$$\mathfrak{Q}(\mu) = \sup_{\varphi \in \mathfrak{B}^{(d)}(\Omega')} |\langle \mu, \varphi \rangle| \tag{23.1}$$

answers the requirements of the lemma.

Corollary 1. – *Same statement as Lemma* 23.1 *with* $A_{\text{loc}}(\Omega)$ *substituted for* $\mathscr{D}'(\Omega)$.

Indeed, the injection $A_{\text{loc}}(\Omega) \to \mathscr{D}'(\Omega)$ is continuous.

In order to complete the proof of Proposition 23.2, we must show that if the pair $(A_{\text{loc}}(\Omega), A_{\text{loc}}(\Omega))$ is $P(\partial/\partial x)$-convex, then the conditions (a), (b), (c) in Proposition 23.1 hold. Let K be a compact subset of Ω, Ω' a relatively compact open neighborhood of K in Ω, and $\mathfrak{Q}$ the seminorm on $\mathscr{D}'(\Omega)$ (or on $A_{\text{loc}}(\Omega)$) defined by (23.1). Let $\varphi \in \gamma_c^{(d)}(\Omega)$ be arbitrary, except that $\operatorname{supp} P(-\partial/\partial x)\,\varphi \subset K$. Then $P(-\partial/\partial x)\,\varphi$ belongs to $\gamma_c^{(d)}(\Omega')$, hence belongs to $C\mathfrak{B}^{(d)}(\Omega')$ for some $C>0$. We derive from this, that for all distributions μ in Ω, in particular, for all $\mu \in A_{\text{loc}}(\Omega)$,

$$|\langle P(-\partial/\partial x)\,\varphi, \mu\rangle| \leqslant C\mathfrak{Q}(\mu).$$

Our hypothesis implies then that there is a continuous seminorm $\mathfrak{p}$ on $A_{\text{loc}}(\Omega)$ such that $\langle \varphi, \mu\rangle = 0$ for all $\mu \in \operatorname{Ker} \mathfrak{p}$. According to (22.5), there is a compact subset K' of Ω such that $\langle \varphi, \mu\rangle = 0$ for all $\mu \in A_{\text{loc}}(\Omega)$, $\operatorname{supp}\mu \subset \Omega \setminus K'$. By Proposition 22.10, this must be true for all μ, say in $\mathscr{E}'(\Omega \setminus K')$, which implies that $\varphi = 0$ in $\Omega \setminus K'$. This means that Property (b), in Proposition 23.1, holds when $A = \gamma^{(d)}$, hence Property (a), as well as (b) and (c) for any A, hold. Q.E.D.

Now, we may state and prove the existence theorem to which we were headed:

Theorem 23.1. – *The following properties are equivalent:*

(a) *the open set Ω is $P(\partial/\partial x)$-convex;*

(b) $P(\partial/\partial x)\, A_{\text{loc}}(\Omega) = A_{\text{loc}}(\Omega)$;

(c) $P(\partial/\partial x)\, \mathscr{D}'(\Omega) \supset A_{\text{loc}}(\Omega)$.

Proof: We have just seen (Proposition 23.2) that (a) implies that the pair $(A_{\text{loc}}(\Omega), A_{\text{loc}}(\Omega))$ is $P(\partial/\partial x)$-convex. In view of Corollary 1 of Proposition 11.2, we may derive (b) from (a) if we prove that the mapping $P(\partial/\partial x): A_{\text{loc}}(\Omega) \to A_{\text{loc}}(\Omega)$ is semiglobally surjective. As the topology of $A_{\text{loc}}(\Omega)$ is local, it suffices to show that, for every relatively compact open subset Ω' of Ω and every $f \in A_{\text{loc}}(\Omega)$, there is $f_1 \in A_{\text{loc}}(\Omega)$ such that

$$P(\partial/\partial x)\, f_1 = f \text{ in } \Omega'. \tag{23.2}$$

There is $g \in A_c$ which is equal to f in Ω' (Proposition 22.3). On the other hand, there exist a fundamental solution $E \in \mathscr{D}'$ of $P(\partial/\partial x)$ (see TREVES [*1*], Corollary 2 of Theorem 2.6); set $h = E * g$. In virtue of Proposition 22.4, $h \in A_{\text{loc}}$. It suffices to take, as solution f_1 of (23.2), the restriction of h to Ω'.

The implication (b)$\Rightarrow$(c) being trivial, it remains to show that (c)$\Rightarrow$(a). The method of proving this implication goes beyond the framework of Theorem 23.1. It is worthwhile stating and proving a somewhat more general result. As before, let Ω be an open subset of $\mathbf{R}^n$, $\mathbf{P}$ a differential operator with $\mathscr{C}^\infty$ coefficients in Ω. We regard $\mathbf{P}$ as a mapping $\mathscr{D}'(\Omega) \to$

$\mathscr{D}'(\Omega)$ and we denote by ${}^t\mathbf{P}$ its transpose, ${}^t\mathbf{P}: \mathscr{C}_c^\infty(\Omega) \to \mathscr{C}_c^\infty(\Omega)$. We recall that $\mathscr{E}'(\Omega)$ denotes the space of distributions in Ω with compact support.

Lemma 23.2. – *Let F be a space of distributions in Ω with the following properties:*

F is a Frechet space of distributions in Ω and its topology is local (cf. (22.4) & (22.5)); (23.3)

for every open subset Ω' of Ω, $F \cap \mathscr{E}'(\Omega')$ is dense in $\mathscr{E}'(\Omega')$. (23.4)

Under these circumstances, if $\mathbf{P}\mathscr{D}'(\Omega) \supset F$, the following holds:

To every compact set $K \subset \Omega$ there is another compact set $K' \subset \Omega$ such that, for all functions $\varphi \in \mathscr{C}_c^\infty(\Omega)$,

$$\operatorname{supp} {}^t\mathbf{P}\varphi \subset K \quad \textit{implies} \quad \operatorname{supp} \varphi \subset K'.$$

Observe that (23.4) implies that F is dense in $\mathscr{D}'(\Omega)$. In order to complete the proof of Theorem 23.1, the only thing left to be done, once the preceding lemma has been proved, is to check that $F = A_{\text{loc}}(\Omega)$ satisfies Condition (23.4). But this is obvious in view of Proposition 22.10 and of the fact that $\mathscr{C}_c^\infty(\Omega')$ is dense in $\mathscr{E}'(\Omega')$.

Proof of Lemma 23.2. Let K be a compact subset of Ω and set

$$\Phi(K) = \{\varphi \in \mathscr{C}_c^\infty(\Omega);\ \operatorname{supp} {}^t\mathbf{P}\varphi \subset K\}.$$

We put on $\Phi(K)$ the topology defined by the seminorms

$$\varphi \rightsquigarrow \|{}^t\mathbf{P}\varphi\|_k \qquad (k = 0, 1, \ldots;\ \text{see p. 59}).$$

Note that these seminorms are in fact norms, since ${}^t\mathbf{P}$ is injective, as the transpose of a map with dense image. Thus $\Phi(K)$ is a metrizable space. We consider the bilinear functional on $F \times \Phi(K)$, $(\mu, \varphi) \rightsquigarrow \langle \mu, \varphi \rangle$, where the bracket expresses the duality between distributions and test functions in Ω. For fixed φ, $\mu \rightsquigarrow \langle \mu, \varphi \rangle$ is a continuous linear functional on $\mathscr{D}'(\Omega)$, a fortiori on F. Now, fix μ. By hypothesis, there is $v \in \mathscr{D}'(\Omega)$ such that $\mathbf{P}v = \mu$, hence $\langle \mu\ \varphi \rangle = \langle v, {}^t\mathbf{P}\varphi \rangle$. But since $\operatorname{supp} {}^t\mathbf{P}\varphi \subset K$, we have for suitable $C > 0$ and k, integer $\geqslant 0$, and for all $\varphi \in \Phi(K)$,

$$|\langle v, {}^t\mathbf{P}\varphi \rangle| \leqslant C\, \|{}^t\mathbf{P}\varphi\|_k.$$

This shows that the linear functional $\varphi \rightsquigarrow \langle \mu, \varphi \rangle$ is continuous on $\Phi(K)$. We apply the corollary of Lemma A.2 (Appendix A) and reach the conclusion that there is $\mathfrak{p} \in \operatorname{Spec} F$ and an integer $k' \geqslant 0$ such that, for all $\mu \in F$ and all $\varphi \in \Phi(K)$,

$$|\langle \mu, \varphi \rangle| \leqslant \mathfrak{p}(\mu)\, \|{}^t\mathbf{P}\varphi\|_{k'}.$$

As the topology of F is local, there is a compact subset K' of Ω such that $\operatorname{supp}\mu \subset \Omega \setminus K'$ implies $\mathfrak{p}(\mu)=0$. Then Property (23.4) implies that $\operatorname{supp}\varphi \subset K'$. Q.E.D.

We have seen (Proposition 21.3) that there are open sets Ω and polynomials P (non identically zero) such that Ω is not $P(\partial/\partial x)$-convex. The usefulness of Theorem 23.1 will be better appreciated if we show that there exist open sets which are $P(\partial/\partial x)$-convex and, as a matter of fact, if the whole space $\mathbf{R}^n$ is one of them. This is easily done if we use Proposition 23.1:

Proposition 23.3. – *Let P be a polynomial in n variables which does not vanish identically. Then every convex open subset of $\mathbf{R}^n$ is $P(\partial/\partial x)$-convex.*

Proof: It follows from the following consequence of the *theorem of supports* (Lions [1]):

Lemma 23.3. – *If $\varphi \in \mathscr{C}_c^\infty(\mathbf{R}^n)$, the convex hull of* $\operatorname{supp}\varphi$ *and the one of* $\operatorname{supp} P(-\partial/\partial x)\,\varphi$ *are equal.*

For a direct, and easy, proof of Lemma 23.3, see Hörmander [1], Lemma 3.4.3, also Treves [1], Theorem 2.7 (cf. loc. cit., Theorem 6.1).

Chapter VIII. Existence and Approximation of Solutions to a Linear Partial Differential Equation

I. General differential operators

24. Approximation of solutions to the homogeneous equation by $\mathscr{C}^\infty$ solutions

In this section, we denote by Ω an open subset of the Euclidean space $\mathbf{R}^n$ and by $\mathbf{P}$, a differential operator (i.e., a linear partial differential operator with $\mathscr{C}^\infty$ coefficients) in Ω. However, all the statements and results extend, with appropriate but obvious adaptations (in particular, concerning duality), to the case where Ω is a $\mathscr{C}^\infty$ manifold, countable at infinity, and $\mathbf{P}$ a V–W-differential operator in Ω, V, W denoting two finite dimensional complex vector bundles over Ω.

In the present section, we begin by considering the following problem: find sufficient conditions in order that, given some space of distributions in Ω, $\mathscr{L}$, every solution u of the homogeneous equation

$$\mathbf{P}u = 0, \tag{24.1}$$

which belongs to $\mathscr{L}$, is the limit, in $\mathscr{L}$, of solutions of the same equation which belong to $\mathscr{C}^\infty(\Omega)$.

We recall what is a space of distributions in Ω: it is a vector subspace of $\mathscr{D}'(\Omega)$ equipped with a locally convex topology finer than (which means that it can be identical to) the topology induced by $\mathscr{D}'(\Omega)$. A space of distributions $\mathscr{L}$ in Ω is said to be *normal* if it contains $\mathscr{C}_c^\infty(\Omega)$, if the natural injection

$$\mathscr{C}_c^\infty(\Omega) \to \mathscr{L} \tag{24.2}$$

is continuous, and if this injection has a dense image. In this case, by transposing the map (24.2), we obtain a continuous injection of $\mathscr{L}'$ into $\mathscr{D}'(\Omega)$. We may express this fact by saying that *the dual of a normal space of distributions is a space of distributions*. Most spaces of distributions are normal: $\mathscr{D}'(\Omega)$, $\mathscr{E}'(\Omega)$, $\mathscr{C}^\infty(\Omega)$, $H^s_{\text{loc}}(\Omega)$ $(s\in\mathbf{R})$, $L^p(\Omega)$ $(1\leqslant p<+\infty)$ are normal. One conspicuous exception: $L^\infty(\Omega)$; this exception shows, among other things, that the dual of a normal space of distributions is not necessarily normal.

In our situation, we shall assume not only that $\mathscr{L}$ is normal, but even that $\mathscr{C}^\infty(\Omega)$ is continuously embedded and dense in $\mathscr{L}$. By transposing the natural injection $\mathscr{C}^\infty(\Omega)\to\mathscr{L}$, we see that $\mathscr{L}'\subset\mathscr{E}'(\Omega)$: $\mathscr{L}'$ is a space of distributions with compact support in Ω.

We obtain a criterion of the desired kind by applying Theorem 17.2:

Theorem 24.1. – *We suppose that* $\mathbf{P}\mathscr{C}^\infty(\Omega)=\mathscr{C}^\infty(\Omega)$. *Let* $\mathscr{L}$ *be a space of distributions in* Ω *such that* $\mathscr{C}^\infty(\Omega)$ *is continuously embedded and dense in* $\mathscr{L}$.

Assume that there is a space of distributions in Ω, $\mathscr{M}$, *having the following properties*:

$\mathscr{M}$ *is barrelled*; (24.3)

$\mathscr{C}^\infty(\Omega)$ *is continuously embedded in* $\mathscr{M}$ *and every point of* $\mathscr{M}$ *belongs to the closure of a subset of* $\mathscr{C}^\infty(\Omega)$ *which is bounded in* $\mathscr{M}$; (24.4)

for every $u\in\mathscr{L}$ *such that* $\mathbf{P}u=0$, *there is a filter* $\mathscr{F}$ *on* $\mathscr{C}^\infty(\Omega)$ *converging to* u *in* $\mathscr{L}$ *and such that* $\mathbf{P}\mathscr{F}$ *converges to zero in* $\mathscr{M}$; (24.5)

for every distribution $\mu\in\mathscr{E}'(\Omega)$, i.e. *with compact support in* Ω, (24.6)

$$ {}^t\mathbf{P}\mu\in\mathscr{L}' \quad \textit{implies} \quad \mu\in\mathscr{M}'. $$

Under these hypotheses, every solution of Eq. (24.1) *belonging to* $\mathscr{L}$ *is the limit of solutions of the same equation belonging to* $\mathscr{C}^\infty(\Omega)$.

Proof: We consider the following commutative diagram

$$ \begin{array}{ccccc} \mathscr{C}^\infty(\Omega) & \xrightarrow{i} & \mathscr{L} & & \\ {\scriptstyle \mathbf{P}_0(=\mathbf{P})}\downarrow & & & \searrow{\scriptstyle \mathbf{P}} & \\ \mathscr{C}^\infty(\Omega) & \underset{j}{\rightarrow} & \mathscr{M} & \underset{k}{\rightarrow} & \mathscr{D}'(\Omega). \end{array} \tag{24.7} $$

The mappings i, j, k are the natural injections. The mapping $\mathbf{P}_0$ is essentially univalent with respect to the pair (i,j) (Definition 13.1 and Proposition 13.1. Hypotheses (24.3) and (24.4) imply (17.3) & (17.4). Since $\mathbf{P}_0$ is surjective, we have (Theorem 17.1) that $(\operatorname{Ker}\mathbf{P}_0)^\perp=\operatorname{Im}{}^t\mathbf{P}_0$. Hence, Condition (b) in Theorem 17.2 reads, in the present context: for every $\lambda\in\mathscr{L}'$, if $\lambda\in\operatorname{Im}{}^t\mathbf{P}_0$, then there is $v\in\mathscr{M}'$ such that $\lambda={}^t\mathbf{P}v$. Evidently, this can be rephrased as follows: if $\mu\in\mathscr{E}'(\Omega)$ verifies ${}^t\mathbf{P}\mu=\lambda$ for some $\lambda\in\mathscr{L}'$, then μ verifies ${}^t\mathbf{P}\mu = {}^t\mathbf{P}\nu$ for some $\nu \in \mathcal{M}'$; as ${}^t\mathbf{P}$ is injective in $\mathscr{E}'(\Omega)$, this means that $\mu = \nu$, i.e., $\mu \in \mathcal{M}'$. In other words, Condition (24.6) is Condition (b). From Theorem 17.2 we derive that the canonical extension (Definition 13.2) of $\mathbf{P}_0$ has the homogeneous approximation property. At this stage, we use Hypothesis (24.5): it says that the kernel of the canonical extension of $\mathbf{P}_0$ is identical to $\operatorname{Ker}\mathbf{P}$ ($\mathbf{P}$ is viewed as a continuous linear map of $\mathscr{L}$ into $\mathscr{D}'(\Omega)$). Then every element of $\operatorname{Ker}\mathbf{P}$ can be approximated by elements of $\operatorname{Ker}\mathbf{P}$ which belong to $\mathscr{C}^\infty(\Omega)$ (this in the sense of the graph topology: but on $\operatorname{Ker}\mathbf{P}$ it is the same as the topology induced by $\mathscr{L}$). Q.E.D.

We shall give right away an application of Theorem 24.1 to linear PDE with constant coefficients. For this, we introduce the following definition:

Definition 24.1. – *We say that a space of distributions in Ω, $\mathscr{L}$, has the property of approximation by cutting and smoothing if there are two sequences in $\mathscr{C}_c^\infty(\Omega)$ and $\mathscr{C}_c^\infty(\mathbf{R}^n)$ respectively, $\{g_j\}$, $\{\varrho_j\}$ $(j=0, 1, \ldots)$ such that:*

(i) *to every compact set $K \subset \Omega$, there is $j_K \geqslant 0$ such that $g_j = 1$ on K for all $j \geqslant j_K$;*

(ii) *to every neighborhood U of 0 in $\mathbf{R}^n$ there is $j_U \geqslant 0$ such that $\operatorname{supp}\varrho_j \subset U$ for all $j \geqslant j_U$;*

(iii) *for every $j = 0, 1, \ldots,$*

$$\operatorname{supp} g_j + \operatorname{supp} \varrho_j \subset \Omega;$$

and such that every element μ of $\mathscr{L}$ be the limit, in $\mathscr{L}$, of the sequence of functions

$$\varrho_j * (g_j \mu), \qquad j = 0, 1, \ldots.$$

If a space $\mathscr{L}$ has the property of approximation by cutting and smoothing, and if $\mathscr{L}$ contains $\mathscr{C}_c^\infty(\Omega)$, $\mathscr{L}$ is obviously normal. All the known normal spaces have the property of approximation by cutting and smoothing: $\mathscr{D}'(\Omega)$, $\mathscr{E}'(\Omega)$, $H^s_{\text{loc}}(\Omega)$, $H^s(\Omega)$ $(s \in \mathbf{R})$, $L^p(\Omega)$ $(1 \leqslant p < +\infty)$, etc. From Theorem 24.1 we derive now:

Theorem 24.2. – *Let $P(\partial/\partial x)$ be a differential operator with constant coefficients in $\mathbf{R}^n$, and Ω an open subset of $\mathbf{R}^n$ which is $P(\partial/\partial x)$-convex.*

Let $\mathscr{L}$ be any space of distributions in Ω, in which $\mathscr{C}^\infty(\Omega)$ is continuously embedded, and which has the property of approximation by cutting and smoothing.

Every solution of the homogeneous equation

$$P(\partial/\partial x)\, u = 0$$

which belongs to $\mathscr{L}$, is the limit, in $\mathscr{L}$, of solutions of the same equation which belong to $\mathscr{C}^\infty(\Omega)$.

Proof: To say that Ω is $P(\partial/\partial x)$-convex (Definition 20.1 and Proposition 23.1) is to say that $P(\partial/\partial x)\,\mathscr{C}^\infty(\Omega)=\mathscr{C}^\infty(\Omega)$ (Theorem 23.1). We shall then apply Theorem 24.1 with $\mathscr{M}=\mathscr{C}^\infty(\Omega)$. Conditions (24.3), (24.4) and (24.6) are trivially satisfied. It remains to show that (24.5) holds also. Let g_j and ϱ_j be functions like the ones considered in Definition 24.1 and let $u\in\mathscr{L}$ verify $P(\partial/\partial x)\,u=0$. Set $u_j=\varrho_j*(g_j u)$. Let Ω' be an arbitrary relatively compact open subset of Ω and U a neighborhood of 0 in $\mathbf{R}^n$ such that $\overline{\Omega'+U}$ is compact and contained in Ω. Let j_0 be large enough to have, for all $j\geqslant j_0$, $\operatorname{supp}\varrho_j\subset U$ and $g_j=1$ in some neighborhood of $\overline{\Omega'+U}$. Then $P(\partial/\partial x)\,u_j=0$ in Ω'. This shows that the functions $P(\partial/\partial x)\,u_j$ converge to 0 in $\mathscr{C}^\infty(\Omega)$. Q.E.D.

Let us give another application of Theorem 24.1, this one valid for a general differential operator $\mathbf{P}$ (with $\mathscr{C}^\infty$ coefficients not necessarily constant) in Ω:

Theorem 24.3. – *Suppose that $\mathbf{P}\mathscr{C}^\infty(\Omega)=\mathscr{C}^\infty(\Omega)$ and that the following condition is satisfied:*

$$\text{For every distribution } \mu \text{ with compact support in } \Omega, \text{ if } {}^t\mathbf{P}\mu \text{ is a } \mathscr{C}^\infty \text{ function, then also } \mu \text{ is a } \mathscr{C}^\infty \text{ function.} \tag{24.8}$$

Then every distribution u in Ω which verifies $\mathbf{P}u=0$ is a limit of solutions of the same equation which belong to $\mathscr{C}^\infty(\Omega)$.

In this statement, "limit" must be understood in the sense of $\mathscr{D}'(\Omega)$.

Proof: Apply Theorem 24.1 with $\mathscr{L}=\mathscr{M}=\mathscr{D}'(\Omega)$.

Examples of differential operators having Property (24.8):

Example 24.1. – The constant coefficients operators. If $P(\partial/\partial x)$ is such an operator, let E be one of its fundamental solutions. Then, for every $\mu\in\mathscr{E}'(\mathbf{R}^n)$, $\mu=E*P(\partial/\partial x)\,\mu$ is a $\mathscr{C}^\infty$ function as soon as this is true of $P(\partial/\partial x)\,\mu$.

Example 24.2. – The hypoelliptic operators (Definition 21.2), in particular the elliptic operators, also the parabolic ones.

Example 24.3. – The strongly hyperbolic operators, more generally the differential operators called *principally normal* by HÖRMANDER ([*1*], Definition 8.5.1 & Theorem 8.7.1; for a general principally normal operator, one must assume that there exists a $\mathscr{C}^2$ function ψ in Ω, such that grad ψ never vanishes in Ω and with pseudoconvex level surfaces throughout Ω, *loc. cit.* Definition 8.6.1; when the operator is strongly hyperbolic in the direction x_n, one may take $\psi(x)=x_n$ – in this case, the open set Ω can be arbitrary).

25. Existence and approximation of solutions to the inhomogeneous equation

We continue to deal with Diagram (24.7):

$$\begin{array}{ccccc} \mathscr{C}^\infty(\Omega) & \xrightarrow{i} & \mathscr{L} & & \\ {\scriptstyle P_0 (=P)}\downarrow & & & \searrow {\scriptstyle P} & \\ \mathscr{C}^\infty(\Omega) & \xrightarrow[j]{} & \mathscr{M} & \xrightarrow[k]{} & \mathscr{D}'(\Omega), \end{array} \tag{25.1}$$

but where now $\mathscr{L}$ and $\mathscr{M}$ are given spaces of distributions in Ω (in Section 24, $\mathscr{L}$ was given and the existence of a space $\mathscr{M}$ with suitable properties was a property of $\mathscr{L}$ which we assumed). We continue to make the assumption:

$\mathscr{C}^\infty(\Omega)$ is continuously embedded and dense in both $\mathscr{L}$ and $\mathscr{M}$; (25.2)

but furthermore:

both $\mathscr{L}$ and $\mathscr{M}$ are Fréchet spaces. (25.3)

Then, the following result is a straightforward consequence of Corollary 3 of Theorem 17.2; we leave its proof to the student:

Theorem 25.1. – *Consider Diagram* (25.1) *and assume that* (25.2) *and* (25.5) *hold. Suppose furthermore that*

$$\mathbf{P}\mathscr{C}^\infty(\Omega) = \mathscr{C}^\infty(\Omega).$$

Then the following properties are equivalent:

(a) *The canonical extension of* $\mathbf{P}: \mathscr{C}^\infty(\Omega) \to \mathscr{C}^\infty(\Omega)$ *with respect to the pair* (i, j) (Definition 13.2) *is an epimorphism and has the homogeneous approximation property* (Definition 14.1);

(b) *for every* $\mu \in \mathscr{E}'(\Omega)$,

$$^t\mathbf{P}\mu \in \mathscr{L}' \quad \textit{implies} \quad \mu \in \mathscr{M}'.$$

Notice that Property (a) implies, among other things, that

$$\mathbf{P}\mathscr{L} \supset \mathscr{M}.$$

We shall mention a few simple applications of Theorem 25.1.

We shall be concerned with the spaces $H^s_{\text{loc}}(\Omega)$ (see Treves [*1*], Section 0.6). Here s is a real number and $H^s_{\text{loc}}(\Omega)$ is the space of distributions u in Ω such that $\varphi u \in H^s$ (see p. 59) for all $\varphi \in \mathscr{C}^\infty_c(\Omega)$. We consider then an arbitrary sequence $\{g_j\}$ ($j=0, 1, \ldots$) in $\mathscr{C}^\infty_c(\Omega)$ such that the set Ω^j, interior of the set $\{x \in \Omega; g_j(x) = 1\}$, contains $\operatorname{supp} g_{j-1}$ for all $j \geq 1$, and such that the union of the Ω^j for all j, is equal to Ω. We provide $H^s_{\text{loc}}(\Omega)$

with the locally convex topology defined by the seminorms

$$u \rightsquigarrow \|g_j u\|_s \qquad (j=0,1,\ldots).$$

This turns $H^s_{\mathrm{loc}}(\Omega)$ into a Fréchet space; it is not difficult to see that it has the property of approximation by cutting and smoothing (Definition 24.1), in particular it is normal (Definition B.2, Appendix B). Its dual is the space $H_c^{-s}(\Omega)$ of elements of H^{-s} which have a compact support contained in Ω. We shall not make use of a topology on $H_c^{-s}(\Omega)$.

The space $\mathscr{C}^\infty(\Omega)$ is continuously embedded, and of course dense, in every space $H^s_{\mathrm{loc}}(\Omega)$; in fact, it is the intersection, as a topological vector space, of the spaces $H^s_{\mathrm{loc}}(\Omega)$ when $s\to+\infty$ (TREVES [1], Theorem 0.52).

When $\mathscr{L}=H^s_{\mathrm{loc}}(\Omega)$, $\mathscr{M}=H^t_{\mathrm{loc}}(\Omega)$ ($s,t\in\mathbf{R}$), Condition (b) in Theorem 25.1 reads:

$$\text{For every } \mu\in\mathscr{E}'(\Omega), \quad {}^t\mathbf{P}\mu\in H^{-s} \text{ implies } \mu\in H^{-t}. \tag{25.4}$$

Examples of operators having Property (25.4)

Example 25.1. – The differential operators with constant coefficients, taking $s=t$. Let $P(\partial/\partial x)$ be such an operator; for all $\mu\in\mathscr{E}'(\mathbf{R}^n)$ (here Ω is obviously irrelevant), $P(-\partial/\partial x)\,\mu\in H^{-s}$ implies $\mu\in H^{-s}$. This follows at once from the result, due to MALGRANGE (see TREVES [1], Theorem 3.2), that every differential operator with constant coefficients (here $P(-\partial/\partial x)$) has a fundamental solution E such that $E*H_c^0\subset H^0_{\mathrm{loc}}$ (it is not difficult to go from the exponent 0 to the exponent $-s$).

Example 25.2. – The elliptic differential operators of order m (with $\mathscr{C}^\infty$ coefficients), taking $s=t+m$. This is a particular case of a general result; for a proof see HÖRMANDER [1], Theorem 7.4.1, Example p. 177.

Example 25.3. – The strongly hyperbolic differential operators of order m, taking $s=t+m-1$. We assume that the operator is hyperbolic in the direction x_n in the whole of Ω; this means in particular that the coefficient of $(\partial/\partial x_n)^m$ never vanishes in Ω. For the proof of the fact that (25.4) holds, see for example HÖRMANDER [1], Theorem 8.7.1 (in this theorem, take $\psi(x)=x_n$). The same theorem provides other examples of operators for which (25.4) holds with $s=t+m-1$.

26. P-Runge domains and relative P-convexity

We consider now two open subsets of $\mathbf{R}^n$, $\Omega\supset\Omega'$, and a differential operator $\mathbf{P}$ on the larger one, Ω. But let us repeat that there is no reason to limit ourselves to this situation: we could as well suppose that Ω is a $\mathscr{C}^\infty$ manifold carrying two vector bundles V, W, Ω' an open subset of Ω and $\mathbf{P}$ a V–W-differential operator in Ω.

Definition 26.1. – *We say that Ω' is a* **P**-*Runge domain of Ω if every solution, belonging to $\mathscr{C}^\infty(\Omega')$, of the homogeneous equation, in Ω',*

$$\mathbf{P}u = 0, \tag{26.1}$$

is the limit, in $\mathscr{C}^\infty(\Omega')$, of solutions of the same equation, in Ω, belonging to $\mathscr{C}^\infty(\Omega)$.

We shall sometimes refer to the pair (Ω, Ω') as a *Runge pair* with respect to the differential operator **P**.

We introduce now the important definition

Definition 26.2. – *We say that Ω' is relatively* **P**-*convex in Ω if, for every distribution μ with compact support in Ω,*

$$\operatorname{supp} {}^t\mathbf{P}\mu \subset \Omega' \quad \textit{implies} \quad \operatorname{supp} \mu \subset \Omega' .$$

Before looking closer at this concept, let us state and prove the result which justifies its introduction:

Theorem 26.1. – *If $\mathbf{P}\mathscr{C}^\infty(\Omega)=\mathscr{C}^\infty(\Omega)$, the following two properties are equivalent:*

(a) *$\mathbf{P}\mathscr{C}^\infty(\Omega')=\mathscr{C}^\infty(\Omega')$ and Ω' is a* **P**-*Runge domain of Ω;*

(b) *Ω' is relatively* **P**-*convex in Ω.*

Proof: Follows at once from Corollary 3 of Theorem 17.2 where we take:

$$E_0 = F_0 = \mathscr{C}^\infty(\Omega), \quad E = F = \mathscr{C}^\infty(\Omega'),$$

$$\varrho = \sigma = \text{restriction to } \Omega' \text{ of functions in } \Omega,$$

$u_0 = \mathbf{P}: \mathscr{C}^\infty(\Omega) \to \mathscr{C}^\infty(\Omega)$ (obviously essentially univalent, Definition 13.1; the canonical extension of u_0 is then $\tilde{u}_0 = \mathbf{P}: \mathscr{C}^\infty(\Omega') \to \mathscr{C}^\infty(\Omega')$, see Definition 13.2).

From Proposition 20.3, we derive the following:

Proposition 26.1. – *Suppose that $\mathbf{P}\mathscr{C}^\infty(\Omega)=\mathscr{C}^\infty(\Omega)$. Then the following properties are equivalent:*

(a) *Ω' is relatively* **P**-*convex in Ω;*

(b) *to every compact subset K of Ω and to every real number s, there is another compact subset K' of Ω' such that, for all $\mu \in \mathscr{E}'(\Omega)$,*

$${}^t\mathbf{P}\mu \in H^s(K) \quad \textit{implies} \quad \operatorname{supp} \mu \subset K' .$$

Proof: (b) implies trivially (a). Conversely, under our assumption, the set $\Gamma^s(K; \mathbf{P}, \Omega)$ is compact (Proposition 20.3) and, in view of (a), it is contained in Ω', whence (b).

We show next that the concept of relative **P**-convexity is strictly stronger than the concept of **P**-convexity. For this, we introduce the following class of differential operators:

Definition 26.3. – *A differential operator* **P** *in* Ω *is said to be analytic-hypoelliptic if every distribution* T *in* Ω *is an analytic function in any open subset of* Ω *where this is true of* $\mathbf{P}T$.

Examples of analytic-hypoelliptic operators

Example 26.1. – The elliptic operators with analytic coefficients (see HÖRMANDER [1], Theorem 7.5.1).

Example 26.2. – The first-order operators $(\partial/\partial x_1)+ix_1^{2k}(\partial/\partial x_2)$ $(k=0, 1,\dots)$ *in* $\mathbf{R}^2$ (MIZOHATA [1]).

Proposition 26.2. – *Let* **P** *be an analytic-hypoelliptic differential operator in the open set* Ω.

Let K *be a compact subset of* Ω, $\hat{K}_\Omega$ *the union of* K *with all the relatively compact connected components of* $\Omega \setminus K$.

The set $\hat{K}_\Omega$ *contains all the sets* $\operatorname{supp}\mu$, *for* $\mu \in \mathscr{E}'(\Omega)$ *such that* $\operatorname{supp}\mathbf{P}\mu \subset K$, *and is equal to their union if* $\mathbf{P}1=0$.

We have denoted by 1 the function identically equal to 1.

Proof: Consider $\mu \in \mathscr{E}'(\Omega)$, $\operatorname{supp}\mathbf{P}\mu \subset K$, and a connected component $\mathcal{O}$ of $\Omega \setminus K$ which is not relatively compact. Then μ verifies $\mathbf{P}\mu=0$ in $\mathcal{O}$, hence is an analytic function in $\mathcal{O}$; but $\operatorname{supp}\mu$ is compact, hence $\mu=0$ in some nonempty open subset of $\mathcal{O}$, hence in the whole of $\mathcal{O}$. Thus $\operatorname{supp}\mu \subset \hat{K}_\Omega$.

Conversely, let $\mathcal{O}$ be a relatively compact connected component of $\Omega \setminus K$, and let μ be its characteristic function. We have $\overline{\mathcal{O}} = \operatorname{supp}\mu$, and if $\mathbf{P}1=0$, $\operatorname{supp}\mathbf{P}\mu \subset K$. Q.E.D.

Remark 26.1. – We can remove the restriction that $\mathbf{P}1=0$ if we know that the equation $\mathbf{P}u=v$ is semiglobally solvable in $\mathscr{C}^\infty(\Omega)$ and replace it by the requirement that the order of **P** is >0. This is easy to check.

Corollary 1. – *If* **P** *is analytic-hypoelliptic in* Ω, Ω *is* ${}^t\mathbf{P}$*-convex.*

Indeed, $\hat{K}_\Omega$ is a compact subset of Ω.

Let then **P** be an analytic-hypoelliptic operator in Ω, and let Ω' be an open subset of Ω such that, for some compact set $K \subset \Omega'$, we have $\hat{K}_\Omega \not\subset \Omega'$. If $\mathbf{P}1=0$, Ω' will not be relatively ${}^t\mathbf{P}$-convex in Ω (while it is ${}^t\mathbf{P}$-convex by Corollary 1). Standard example: $\mathbf{P}=(\partial/\partial x_1)+i(\partial/\partial x_2)$ in $\Omega=\mathbf{R}^2$, $\Omega'=\{(x_1, x_2);\ x_1^2+x_2^2 \neq 0\}$.

The case of constant coefficients differential operators

Let P be a polynomial in n variables, with complex coefficients which are not all equal to zero, $\Omega' \subset \Omega$ two open subsets of $\mathbf{R}^n$. It is easy to see that Ω' is relatively $P(\partial/\partial x)$-convex in Ω if and only if, for every function $\varphi \in \mathscr{C}_c^\infty(\Omega)$,

$$\operatorname{supp} P(-\partial/\partial x)\,\varphi \subset \Omega, \quad \text{implies} \quad \operatorname{supp} \varphi \subset \Omega,.$$

(cf. Proposition 23.1). Then it follows at once from Lemma 23.3 that, if Ω' is *convex*, Ω' is relatively $P(\partial/\partial x)$-convex in Ω whatever be P and Ω.

Chapter IX. Existence and Approximation of Solutions to a Linear Partial Differential Equation

II. Differential operators with constant coefficients

27. Spaces of polynomials, of formal power series, of exponential-polynomials, of entire functions of exponential type

We shall denote by $\mathscr{P}_n$ the vector space of polynomials in n indeterminates (or "variables") $X_1, \dots, X_n$, with complex coefficients, by $\mathscr{P}_n^d (d=0, 1, \dots)$ the subspace of $\mathscr{P}_n$ consisting of the polynomials of degree $\leqslant d$. We provide $\mathscr{P}_n$ with the locally convex topology, inductive limit of the finite-dimensional Hausdorff TVS $\mathscr{P}_n^d$: a seminorm $\mathfrak{p}$ on $\mathscr{P}_n$ is continuous if and only if its restriction to every $\mathscr{P}_n^d$ for $d=0, 1, \dots$, is continuous. But every seminorm on a finite dimensional Hausdorff TVS is continuous (cf. Treves [2], Theorem 9.1), therefore this is also true in $\mathscr{P}_n$: Spec $\mathscr{P}_n$ is the total spectrum of $\mathscr{P}_n$, in other words, *$\mathscr{P}_n$ carries the finest locally convex topology*. This is equivalent with saying that every linear functional on $\mathscr{P}_n$ is continuous: *the algebraic dual of $\mathscr{P}_n$ is equal to its dual.*

We shall denote by $\mathscr{Q}_n$ the vector space of formal power series, with complex coefficients, in n indeterminates. We provide it with the topology of convergence of each coefficient: a sequence of formal power series,

$$u = \sum_{p \in \mathbf{N}^n} u_p X \qquad (X^p = X_1^{p_1} \dots X_n^{p_n}),$$

converges to zero if, for each p, the numbers u_p converge to 0. It is seen at once that $\mathscr{Q}_n$ is a Fréchet space. Also every bounded subset of $\mathscr{Q}_n$ has a compact closure, hence $\mathscr{Q}_n$ is a Montel space (Treves [2], Example 36.4), in particular it is reflexive (loc. cit., Definition 36.2).

The TVS $\mathscr{P}_n$ and $\mathscr{Q}_n$ can be put into duality by the bracket

$$(P, u) \rightsquigarrow \langle P, u \rangle = \sum_{p \in \mathbf{N}^n} \frac{1}{p!} P^{(p)}(0)\, u^{(p)}(0), \tag{27.1}$$

with obvious notation. This turns each space into the dual of the other one, as can easily be proved (TREVES [2], Theorem 22.1). From now on, we identify $\mathscr{P}_n$ and $\mathscr{Q}'_n$, $\mathscr{Q}_n$ and $\mathscr{P}'_n$.

We call *exponential-polynomial* function in a vector space E any function of the type

$$x \rightsquigarrow f(x) = \sum_{x'} f(x, x')\, e^{\langle x', x \rangle},$$

where x' varies in the dual E' of E, where, for each x', $x \rightsquigarrow f(x, x')$ is a polynomial function in E with complex values, identically equal to zero for all x' except possibly a finite number of them. We shall be interested in the case where $E = \mathbf{C}^n$. We see, in this case, that an exponential-polynomial function is an element of the direct sum

$$EP = \coprod_{z \in \mathbf{C}^n} e^{\langle z, x \rangle} \mathscr{P}_n,$$

where we have identified $\mathbf{C}^n$ with its own dual, by using the coordinates $x_1, \ldots, x_n$. Each "factor" $e^{\langle z, x \rangle} \mathscr{P}_n$ is a copy of $\mathscr{P}_n$ via the multiplication by $\exp(\langle z, x \rangle)$; we carry on it the topology of $\mathscr{P}_n$. We may now define a locally convex topology on EP: a seminorm on EP is continuous if and only if its restriction to each $e^{\langle z, x \rangle} \mathscr{P}_n$ is continuous; but on the latter, all seminorms are continuous, therefore this is also true on EP.

The dual of EP, which is also the algebraic dual of EP, is the product

$$(EP)' = \prod_{z \in \mathbf{C}^n} (e^{\langle z, x \rangle} \mathscr{P}_n)'.$$

This dual EP' may be identified with the product $\prod_{z \in \mathbf{C}^n} (\mathscr{Q}_n)_z$, where $(\mathscr{Q}_n)_z$ is a copy of $\mathscr{Q}_n$, and the duality with EP is defined by the following bracket: if $\mu = (u_z)_{z \in \mathbf{C}^n}$ and $f = \sum_{z \in \mathbf{C}^n} f_z\, e^{\langle x, z \rangle}$, $u_z \in \mathscr{Q}_n$, $f_z \in \mathscr{P}_n$ for all z, and $f_z \equiv 0$ for all z except possibly a finite number of them,

$$\langle \mu, f \rangle = \sum_z \langle u_z, f_z \rangle, \tag{27.2}$$

where the brackets in the sum are those of the duality between $\mathscr{P}_n$ and $\mathscr{Q}_n$.

We shall denote by Exp the space of entire functions in $\mathbf{C}^n$ of exponential type, that is to say of the functions which satisfy for suitable (finite!) constants A, $B > 0$,

$$\textit{for all } z \in \mathbf{C}^n, \ |f(z)| \leqslant A\, e^{B|z|}. \tag{27.3}$$

If $f \in \mathrm{Exp}$, the infimum of the numbers B such that (27.3) holds is called the *exponential type* of f. For a given B, let us denote by $\mathrm{Exp}(B)$ the space of entire functions which satisfy (27.3) for some $A > 0$ (this is not the same as saying "the entire functions of exponential type B"!). If $f \in \operatorname{Exp}(B)$, the infimum of the numbers A such that (27.3) holds, is a norm; with this norm, $\operatorname{Exp}(B)$ is a Banach space. If $B<B'$, we have the natural injection of $\operatorname{Exp}(B)$ into $\operatorname{Exp}(B')$. It is obviously continuous, but it is *not* a monomorphism, i.e., the norm in $\operatorname{Exp}(B)$ is not equivalent with the one induced by $\operatorname{Exp}(B')$. At any event, we put on Exp the locally convex topology, inductive limit of the $\operatorname{Exp}(B)$, as $B \to +\infty$.

Let now H, or $H(\mathbf{C}^n)$ if precision is needed, denote the space of entire functions in $\mathbf{C}^n$, equipped with the topology of uniform convergence on the compact subsets of $\mathbf{C}^n$. The space H is a Fréchet and a Montel space.

The spaces H and Exp can be regarded as the dual of each other via the bracket

$$(h, f) \rightsquigarrow \langle h, f \rangle = \sum_{p \in \mathbf{N}^n} \frac{1}{p!} h^{(p)}(0)\, f^{(p)}(0). \tag{27.4}$$

This assertion is not difficult to prove (see e.g., TREVES [2], Theorem 22.2).

Finally, we see that we are dealing with the following two sequences of "natural mappings":

$$\mathscr{P}_n \to EP = \coprod_{z \in \mathbf{C}^n} e^{\langle x, z \rangle} \mathscr{P}_n \to \operatorname{Exp} \to H \to \mathscr{Q}_n, \tag{27.5}$$

$$\mathscr{Q}_n \leftarrow (EP)' = \prod_{z \in \mathbf{C}^n} (\mathscr{Q}_n)_z \quad \leftarrow \quad H \leftarrow \operatorname{Exp} \leftarrow \mathscr{P}_n. \tag{27.6}$$

The first three arrows in (27.5) are the natural injections; the last one is the assignement, to any entire function, of its Taylor expansion about the origin. In (27.6), the map $(EP)' \to \mathscr{Q}_n$ is simply the coordinate projection on $(\mathscr{Q}_n)_{z=0}$. Sequences (27.5) and (27.6) are dual of each other: two spaces in the same column are the dual of each other, two mappings in the same column are the transpose of each other. We shall now look closer at the map $H \to (EP)'$ in (27.6). In the whole sequel, when an entire function is identified with a power series, this is done through its Taylor expansion about the origin.

Let f be a formal power series (resp., an entire function) in the n variables x_j $(1 \leqslant j \leqslant n)$, g a polynomial (resp., an entire function of exponential type) in the same variables, P a second such polynomial. Let the brackets mean the duality brackets between $\mathscr{Q}_n$ and $\mathscr{P}_n$ (resp. H and Exp). I claim that we have

$$\langle f, Pg \rangle = \langle P(\partial/\partial x) f, g \rangle. \tag{27.7}$$

Because of the linearity of (27.7) with respect to P, it suffices to prove the formula with respect to $P(x)=x^p=x_1^{p_1}\dots x_n^{p_n}$, and in view of evident induction on p, only when $P(x)=x_1$. Now,

$$\langle f, x_1 g\rangle = \sum_{p\in\mathbf{N}^n} \frac{1}{p^*!} f^{(p)}(0)\, g^{(p^*)}(0),$$

where $p^*=(p_1-1, p_2, \dots, p_n)$ if $p=(p_1,\dots,p_n)$ and $p_1>0$ (otherwise we agree that $(p^*!)^{-1}=0$). But this means that

$$\langle f, x_1 g\rangle = \sum_{p^*\in\mathbf{N}^n} \frac{1}{p^*!} [(\partial/\partial x_1) f^{(p^*)}(0)]\, g^{(p^*)}(0),$$

which is what we wanted to prove. Because of the obvious symmetry, (27.7) remains true when we exchange $\mathscr{P}_n$ and $\mathscr{Q}_n$, H and Exp.

This being said, we assume that f is an entire function and choose $P(x)=(\langle x, z\rangle)^k$, $z\in\mathbf{C}^n$ and $k\in\mathbf{N}$ arbitrary. Then (27.7) reads

$$\langle f, (\langle x, z\rangle)^k g\rangle = \langle(\langle z, \partial/\partial x\rangle)^k f, g\rangle.$$

If we multiply each one of these inequalities by $1/k!$ for $k=0, 1, \dots$, and add them all, we obtain, with obvious "symbolic" notation,

$$\langle f, e^{\langle x, z\rangle} g\rangle = \langle e^{\langle z, \partial/\partial x\rangle} f, g\rangle.$$

But it is seen at once that

$$e^{\langle z, \partial/\partial x\rangle} f(x) = f(x+z) = \tau_{-z} f(x),$$

the translation of f by $-z$. Finally:

$$\langle f, e^{\langle x, z\rangle} g\rangle = \langle \tau_{-z} f, g\rangle. \tag{27.8}$$

The study of the mapping $H\to(EP)'$ is now easy; this map is the transpose of the natural injection $i: EP\to\text{Exp}$. If $h\in H$, we have, for all elements $\sum_z e^{\langle x, z\rangle} f_z$ of EP,

$$\langle {}^t i h, \sum_z e^{\langle x, z\rangle} f_z\rangle = \sum_z \langle h, e^{\langle x, z\rangle} f_z\rangle,$$

where the bracket, in the left hand side, is the one of the duality between (EP) and $(EP)'$, whereas, in the sum of the right hand side, it is the one of the duality between H and Exp. In view of (27.8), we obtain

$$\langle {}^t i h, \sum_z e^{\langle x, z\rangle} f_z\rangle = \sum_z \langle \tau_{-z} h, f_z\rangle.$$

This means exactly, in view of (27.2), that

$${}^t i h = (\tau_{-z} h)_{z\in\mathbf{C}^n}, \tag{27.9}$$

and describes completely the mapping $H\to(EP)'$ in the sequence (27.6).

28. Existence of solutions in the spaces of polynomials, of formal power series, of exponential-polynomials, of entire functions of exponential type

We consider now a linear partial differential operator with constant (i.e., complex) coefficients, in the n variables $x_1, \ldots, x_n$, which we denote by $P(\partial/\partial x)$. It operates in the natural manner on polynomials, formal power series, entire functions in $\mathbf{C}^n$ (then one may interpret the x_j's as complex variables).

Theorem 28.1. – *We have*

$$P(\partial/\partial x)\,\mathscr{P}_n = \mathscr{P}_n, \quad P(\partial/\partial x)\,\mathscr{Q}_n = \mathscr{Q}_n.$$

Proof: In both cases, the transpose of the mapping $P(\partial/\partial x)$ in the sense of the duality between $\mathscr{P}_n$ and $\mathscr{Q}_n$ is the operator of multiplication by the polynomial $P(x)$ (Formula (27.7)). It is therefore injective. Now, a linear map is surjective if and only if its algebraic transpose is injective; this proves the result for $\mathscr{P}_n$. For $\mathscr{Q}_n$ we apply Theorem 17.1: it suffices to show that, in $\mathscr{P}_n$, the ideal $P\mathscr{P}_n$ is equal to the orthogonal of $(\operatorname{Ker} P(\partial/\partial x))^\perp$. On reduces at once the problem to finite dimensional vector spaces: in these, it is always true that the image of the transpose is equal to the orthogonal of the kernel of the linear map under study.

Corollary 1. – *The ideal $P\mathscr{Q}_n$ is closed in $\mathscr{Q}_n$.*

Proof: As $P(\partial/\partial x): \mathscr{P}_n \to \mathscr{P}_n$ is surjective, it is an epimorphism. Let $N = \operatorname{Ker} P(\partial/\partial x)$ (in $\mathscr{P}_n$); $P(\partial/\partial x)$ defines an isomorphism of $\mathscr{P}_n/N$ onto $\mathscr{P}_n$ whose transpose is an isomorphism of $\mathscr{Q}_n$ onto $N^\perp$; but this transpose is the multiplication by the polynomial P.

Corollary 2. – *$P(\partial/\partial x)\,EP = EP$, the space of exponential-polynomial functions in $\mathbf{C}^n$.*

Indeed, consider $f = \sum_{z \in \mathbf{C}^n} f_z\, e^{\langle x, z\rangle}$ ($f_z \in \mathscr{P}_n$ is equal to zero for all z except possibly a finite number of them). For each z such that $f_z \neq 0$, choose $u_z \in \mathscr{P}_n$ such that

$$P(\partial/\partial x + z)\, u_z = f_z.$$

Then $P(\partial/\partial x)\, u = f$, if $u = \sum_{z \in \mathbf{C}^n} u_z\, e^{\langle x, z\rangle}$.

Let us take a look at the transpose of $P(\partial/\partial x): EP \to EP$ for the duality between EP and $(EP)' = \prod_{z \in \mathbf{C}^n} (\mathscr{Q}_n)_z$. Consider $f \in EP$ as before, and let $g = (g_z)_{z \in \mathbf{C}^n} \in (EP)'$. We have:

$$P(\partial/\partial x) f = \sum_z e^{\langle x, z\rangle}\, P(\partial/\partial x + z)\, f_z.$$

Thus:

$$\langle {}^tP(\partial/\partial x)g, f\rangle = \langle g, P(\partial/\partial x) f\rangle = \sum_z \langle g_z, P(\partial/\partial x + z) f_z\rangle.$$

If we use then (27.7), setting $\tau_{-z}P(z')=P(z'+z)$, we obtain

$$\langle {}^tP(\partial/\partial x)\, g, f\rangle = \sum_z \langle(\tau_{-z}P)\, g_z, f_z\rangle .$$

Finally, we see that

$$ {}^tP(\partial/\partial x)\, g = ((\tau_{-z}P)\, g_z)_{z\in \mathbf{C}^n} . \tag{28.1}$$

In particular, since $(\tau_{-z}P)\, \mathcal{Q}_n$ is closed in $\mathcal{Q}_n$, for every z, in virtue of Corollary 1, we obtain:

Corollary 3. – *The image ${}^tP(\partial/\partial x)\,[(EP)']$ is closed in $(EP)'$.*

Of course, we could have proved directly Corollary 3, exactly as we did for Corollary 1.

Theorem 28.2. – $P(\partial/\partial x)$ Exp $=$ Exp.

Proof: Theorem 28.2 is a consequence of the following lemma:

Lemma 28.1. – *Let P be a polynomial. There is a constant $C>0$ such that, for all entire functions h in $\mathbf{C}^n$ and all $z\in\mathbf{C}^n$,*

$$|h(z)| \leqslant C \sup_{|z-z'|<1} |P(z')\, h(z')| .$$

Derivation of Theorem 28.2 from Lemma 28.1

Lemma 28.1 implies, for every compact set $K\subset\mathbf{C}^n$,

$$\sup_{z\in K} |h(z)| \leqslant C \sup_{d(z',K)<1} |P(z')\, h(z')| .$$

This means that $h\rightsquigarrow Ph$ is a monomorphism of H into itself; its transpose, for the duality between H and Exp, is therefore an epimorphism; this transpose is precisely $P(\partial/\partial x)$, by Formula (27.7).

Proof of Lemma 28.1. – By performing a linear change of variables, we may assume that $P(z)=z_1^m+$terms of degree $\leqslant m-1$ with respect to z_1, and by fixing arbitrarily the variables $z_2,\ldots,z_n$, we see that it suffices to show, for all polynomials P_1 in one variable, z_1, of the form $P_1(z_1)= z_1^m+$terms of degree $\leqslant m-1$, for all entire functions φ in the complex plane, for all $z_1\in\mathbf{C}^1$,

$$|\varphi(z_1)| \leqslant \sup_{|z_1'-z_1|<1} |P_1(z_1')\, \varphi(z_1')| .$$

By a translation and in view of the maximum principle, this is equivalent with

$$|\varphi(0)| \leqslant \sup_{|z_1'|=1} |P_1(z_1')\, \varphi(z_1')| . \tag{28.2}$$

Let $\bar{P}_1$ be the same polynomial as P_1 except that the coefficients have been replaced by their complex conjugates; set

$$Q(t) = t^m \bar{P}_1(t^{-1}).$$

For $|z_1'|=1$, we have $|Q(z_1')|=|P_1(z_1')|$; and $Q(0)=1$. Thus (28.2) is equivalent with

$$|Q(0)\,\varphi(0)| \leqslant \sup_{|z_1'|=1} |Q(z_1')\,\varphi(z_1')|.$$

But this is evident. Q.E.D.

29. Existence and approximation of solutions in the space of entire functions

We keep the notation of the previous Sections. We consider the commutative diagram

$$\begin{array}{ccc} E_0 & \xrightarrow{i} & H \\ {\scriptstyle P(\partial/\partial x)}\downarrow & & \downarrow{\scriptstyle P(\partial/\partial x)} \\ E_0 & \xrightarrow[i]{} & H, \end{array}$$

where E_0 will be successively the space EP of exponential-polynomials and the space $\mathscr{P}_n$ of polynomials; $H=H(\mathbf{C}^n)$, the space of entire functions in $\mathbf{C}^n$. By i we have denoted the natural injection. It is clear that $P(\partial/\partial x)$: $E_0 \to E_0$ is essentially univalent with respect to (i, i) (Definition 13.1) and that $P(\partial/\partial x): H \to H$ is its canonical extension (Definition 13.2). We shall be able to apply Corollary 2 of Theorem 17.2. Indeed, as $P(\partial/\partial x)$ is a surjection of E_0 onto itself (Theorem 28.1 and Corollary 2), and as H is a Fréchet space, Conditions (17.3) and (17.4) are satisfied. On the other hand (Corollaries 1 & 3 of Theorem 28.1), the condition (17.5) is also satisfied; and so is (17.6), again in view of the surjectivity of $P(\partial/\partial x)$: $E_0 \to E_0$. Thus we see that the following two properties are equivalent:

(a) *$P(\partial/\partial x)$ is a homomorphism and its kernel in E_0 is dense in its kernel in H;*

(b) *for every $g \in E_0'$,*

$${}^tP(\partial/\partial x)\, g \in {}^t i(H') \quad \textit{implies} \quad g \in {}^t i(H').$$

We begin by observing that (a) can be rephrased as follows:

(a′) *$P(\partial/\partial x)$ is an epimorphism of H onto itself and its kernel in E_0 is dense in its kernel in H.*

Indeed, $P(\partial/\partial x): H \to H$ is a homomorphism with dense image ($\mathscr{P}_n$ is dense in H), hence an epimorphism (Theorem 17.1).

Next we study (b). Note that ${}^t i$ is the compose of the mappings

$$\mathrm{Exp} \to H \to (EP)'$$

in (27.6). Let $E_0 = EP$. Then, in view of (27.9), for all $h \in \mathrm{Exp}$,

$${}^t i(h) = (\tau_{-z} h)_{z \in \mathbf{C}^n},$$

and from (28.1) we derive that, for any $g \in (EP)'$,

$${}^t P(\partial/\partial x)\, g = ((\tau_{-z} P)\, g_z)_{z \in \mathbf{C}^n}.$$

In view of all this, Property (b) reads as follows when $E_0 = EP$:

(b′) *For any* $(g_z)_{z \in \mathbf{C}^n} \in (EP)'$, *if there is* $h \in \mathrm{Exp}$ *such that, for all* $z \in \mathbf{C}^n$,

$$(\tau_{-z} P)\, g_z = \tau_{-z} h,$$

then there is $u \in \mathrm{Exp}$ *such that, for all* z, $g_z = \tau_{-z} u$.

When $E_0 = \mathscr{P}_n$, it follows from (27.7) that ${}^t P(\partial/\partial x)\, g = Pg$, product of the formal power series g with the polynomial P; on the other hand, ${}^t i(h) = h$, viewed as a power series (equal to the Taylor expansion of h about zero). Thus, when $E_0 = \mathscr{P}_n$, (b) reads:

(b″) *For any* $g \in \mathscr{Q}_n$, *if there is* $h \in \mathrm{Exp}$ *such that*

$$Pg = h,$$

then $g \in \mathrm{Exp}$.

Theorem 29.1. – *Suppose that the polynomial P is not identically equal to zero. Then:*

1) $P(\partial/\partial x)$ *is an epimorphism of* $H(\mathbf{C}^n)$ *onto itself;*

2) *every solution* $h \in H(\mathbf{C}^n)$ *of the homogeneous equation*

$$P(\partial/\partial x)\, h = 0 \tag{29.1}$$

is the limit, in $H(\mathbf{C}^n)$, *of exponential-polynomial solutions to the same equation.*

Proof: It suffices to show that (b′) is true. We apply the following particular case of the Vorbereitung Satz in analytic geometry (for a proof, see TREVES [1], Lemma 4.2):

Lemma 29.1. – *Let u, v, w be three formal power series such that $uv = w$. If two of them are convergent series, the third one is also.*

A formal power series is said to be convergent if it converges in some neighborhood of the origin.

Let then $h \in \text{Exp}$, $(g_z)_{z\in\mathbf{C}^n} \in (EP)'$ be such that, for every z,

$$(\tau_{-z}P)\, g_z = \tau_{-z}h\,.$$

In view of Lemma 29.1, g_z is the Taylor expansion about $x=0$ of a holomorphic function G_z in some neighborhood of 0. Let us set $u_z(x)=G_z(x-z)$ for x in some neighborhood of z. We have

$$Pu_z = h$$

in some open neighborhood of z, $U(z)$. If z' is another point of $\mathbf{C}^n$, u_z and $u_{z'}$ are equal in the set

$$U(z) \cap U(z') \cap \{x \in \mathbf{C}^n;\, P(x) \neq 0\},$$

and therefore also in $U(z) \cap U(z')$. In other words, for all z, u_z is the restriction to $U(z)$ of an entire function u; we have

$$g_z = \tau_{-z}u$$

(i.e., $g_z(x)=u(x+z)$ for all x). It remains to show that u is of exponential type. We know that Pu is exponential type: it suffices therefore to apply Lemma 28.1. Q.E.D.

We examine now the case in which $E_0 = \mathscr{P}_n$. Then it is certainly not sufficient, in order that (a′) hold, that P be nonidentically zero. For in general, (b″) will not be true: take $P(x)=x_1-1$, $h(x)\equiv 1$; then the power series $(x_1-1)^{-1}$ is not an entire function. However, one may say the following:

Theorem 29.2: – *The following conditions are equivalent:*

(a) *every solution* $h \in H(\mathbf{C}^n)$ (*resp.* $h \in \mathscr{Q}_n$) *of* (29.1) *is the limit in* $H(\mathbf{C}^n)$ (*resp. in* $\mathscr{Q}_n$) *of the polynomial solutions to the same equation*;

(b) *every irreducible factor of* P *vanishes at the origin.*

Proof: (a)⇒(b), rather non-(b) implies non-(a). Suppose that $P=QR$ with $\deg Q>0$, $\deg R>0$, and $R(0)\neq 0$; let R^{-1} be the inverse of R in the ring $\mathscr{Q}_n$. We have $QR(R^{-1})=Q$; and R^{-1} is not en entire function.

(b)⇒(a). The proof is based on the following lemma:

Lemma 29.2. – *Let Q be an irreducible polynomial vanishing at the origin, and h an entire function such that h/Q is holomorphic in some neighborhood of* 0. *Then h/Q is an entire function.*

Proof of Lemma 29.2: Let $V=\{x\in\mathbf{C}^n;\, Q(x)=0\}$. As Q is irreducible, V is an irreducible analytic set, which means that the set V^0 of its regular points is connected. The restriction of h to V^0 is holomorphic; as it vanishes in some neighborhood in V of the origin (which lies on V), it vanishes everywhere in V^0. Let $z^0 \in V^0$; by definition of a regular point, we may perform a linear change of variables in $\mathbf{C}^n$ so as to have, in some neighborhood of z^0, $Q(z)=(z_n-z_n^0)\, q(z, z^0)$, with $q(z^0, z^0)\neq 0$. But then, for some integer $k\geqslant 1$, in a neighborhood of z^0,

$$h(z) = (z_n - z_n^0)^k \, H(z, z^0).$$

This implies at once that h/Q is holomorphic in a neighborhood of z^0, i.e., h/Q is holomorphic in $\mathbf{C}^n \smallsetminus V^1$, where $V^1 = V \smallsetminus V^0$. Now, h/Q is a meromorphic function; the set of its poles, if it is not empty, is of (complex) codimension 1; but V^1 is of codimension at least equal to 2, hence h/Q has no poles. Q.E.D.

End of the proof of Theorem 29.2. Write $P=P_1\ldots P_r$ where the P_j are irreducible polynomials of degree >0. Suppose that every P_j vanishes at the origin and let $g\in\mathscr{Q}_n$ satisfy $Pg=h\in\mathrm{Exp}$ (resp. $h\in\mathscr{P}_n$). By Lemma 29.1 we know that h/P_r is holomorphic near the origin; by Lemma 29.2 we see that h/P_r is an entire function. We apply Lemma 28.1: if $h\in\mathrm{Exp}$, we derive that $h/P_r\in\mathrm{Exp}$; if h is a polynomial, we see that h/P_r is of polynomial growth at infinity. In this case, Liouville's theorem tells us that h/P_r must be a polynomial. By induction on r, we reach the same conclusions for g. Q.E.D.

30. Further theorems of existence and approximation of solutions

30.1 *Existence and approximation of solutions in the space $\mathscr{C}^\infty$*

Consider the natural injection of $H=H(\mathbf{C}^n)$ into $\mathscr{C}^\infty=\mathscr{C}^\infty(\mathbf{R}^n)$, that it is to say the restriction to the real space of entire analytic functions in the complex space.

It has a dense image, its transpose is an injection of $\mathscr{E}'$, the space of distributions with compact support in $\mathbf{R}^n$, which is the dual of $\mathscr{C}^\infty$, into Exp, the space of entire functions of exponential type, which we have identified, in the previous sections, with the dual of H. We wish to describe more in detail this injection of $\mathscr{E}'$ into H; we denote it momentarily by J. We have, for all $\mu\in\mathscr{E}'$ and all $h\in H$,

$$\langle\mu, h\rangle = \sum_p \frac{1}{p!} (J\mu)^{(p)}(0)\, h^{(p)}(0).$$

Let us choose $h(x)=\exp(\langle\zeta, x\rangle)$; we obtain:

$$(J\mu)(\zeta) = \langle\mu, \exp(\langle\zeta, x\rangle)\rangle.$$

We see that if we denote by $\hat{\mu}$ the Fourier transform of μ, we have, for all $\zeta \in \mathbf{C}^n$,

$$J\mu(\zeta) = \hat{\mu}(i\zeta) \qquad (i = \sqrt{-1}).$$

If we apply the Paley-Wiener theorem (extended to distributions; see HÖRMANDER [1], Theorem 1.7.7), we see that $f \in \mathrm{Exp}$ belongs to $\mathrm{Im}\, J$ if and only if the restriction of f to the purely imaginary space $i\mathbf{R}^n$ grows at infinity slower than some polynomial. We may therefore state and prove:

Theorem 30.1. – *Let P be a polynomial in n variables, with complex coefficients, non identically zero. Then:*
1) *$P(\partial/\partial x)$ is an epimorphism of $\mathscr{C}^\infty(\mathbf{R}^n)$ onto itself;*
2) *every solution $u \in \mathscr{C}^\infty(\mathbf{R}^n)$ of the homogeneous equation*

$$P(\partial/\partial x)\, u = 0 \tag{30.1}$$

is the limit, in $\mathscr{C}^\infty(\mathbf{R}^n)$, of exponential-polynomial solutions of the same equation.

Proof: In virtue of Theorem 29.1, it suffices to prove 1) and
2′) *every solution $u \in \mathscr{C}^\infty$ of* (30.1) *is the limit, in $\mathscr{C}^\infty$, of solutions of* (30.1) *belonging to H.*

We consider the diagram

$$\begin{array}{ccc} H & \rightarrow & \mathscr{C}^\infty \\ {\scriptstyle P(\partial/\partial x)}\downarrow & & \downarrow{\scriptstyle P(\partial/\partial x)} \\ H & \rightarrow & \mathscr{C}^\infty, \end{array}$$

and apply Corollary 3 of Theorem 17.2. We must prove that Condition (b) in this corollary holds, i.e., that for every $h \in \mathrm{Exp}$,

$$Ph \in J(\mathscr{E}') \text{ implies } h \in J(\mathscr{E}'). \tag{30.2}$$

But in view of what we have said above, (30.2) says that if Ph is "slowly growing at infinity" on $i\mathbf{R}^n$, the same is true of h. This follows at once from Lemma 28.1. Q.E.D.

30.2 *Existence and approximation of solutions in the space $H(\Omega)$ of holomorphic functions in a convex open subset of the complex space*

Let at first Ω be any open subset of $\mathbf{C}^n$, not necessarily convex. The space $H(\Omega)$ is provided with the topology of uniform convergence on the compact subsets of Ω. We shall denote by $H'(\Omega)$ the dual of $H(\Omega)$. The elements of $H'(\Omega)$ are called the *analytic functionals* in Ω. When $\Omega = \mathbf{C}^n$, the duality bracket (27.4) establishes an isomorphism between $H' =$

$H'(\mathbf{C}^n)$ and Exp. This isomorphism assigns to any analytic functional μ on $\mathbf{C}^n$ the function $\hat{\mu} \in \text{Exp}$ such that, for all $h \in H$,

$$\langle \mu, h \rangle = \sum_p \frac{1}{p!} \hat{\mu}^{(p)}(0)\, h^{(p)}(0).$$

Taking $h(x) = \exp\langle x, z\rangle$, we see that

$$\hat{\mu}(z) = \langle u_x, \exp\langle x, z\rangle\rangle.$$

The isomorphism $\mu \rightsquigarrow \hat{\mu}$ is called *Fourier-Borel transformation.*

Let us denote by r_Ω the restriction of entire functions to the open set Ω; the transpose of r_Ω is a mapping of $H'(\Omega)$ into H'. An analytic functional in $\mathbf{C}^n$ is said to be *carried* by Ω if it belongs to $\operatorname{Im}{}^t r_\Omega$. We recall that Ω is called a *Runge domain* if $\operatorname{Im} r_\Omega$ is dense in $H(\Omega)$; then, and only then, ${}^t r_\Omega$ is injective.

Proposition 30.1. – *Let Ω be a Runge domain in $\mathbf{C}^n$. The following properties*

1) $P(\partial/\partial x)\, H(\Omega) = H(\Omega)$,
2) *every solution of the homogeneous equation* (30.1) *in $H(\Omega)$ is the limit, in $H(\Omega)$, of exponential-polynomial solutions to the same equation,*

hold if and only if the following property holds:

for every analytic functional μ in $\mathbf{C}^n$, if ${}^tP(\partial/\partial x)\,\mu$ is carried by Ω, then μ also is carried by Ω. (30.3)

Proof: It suffices to apply Corollary 3 of Theorem 17.2, taking advantage of the fact that $r_\Omega(H)$ is dense in $H(\Omega)$.

Theorem 30.2. – *Let Ω be a convex open subset of $\mathbf{C}^n$. Then*:

1) $P(\partial/\partial x)\, H(\Omega) = H(\Omega)$;
2) *every solution $h \in H(\Omega)$ of* Eq. (30.1) *is the limit, in $H(\Omega)$, of exponential-polynomials solutions of* (30.1).

Proof: A convex open subset of $\mathbf{C}^n$ is a Runge domain, therefore it suffices to prove that (30.3) holds when Ω is convex. The proof of this fact is based on an important result, which we proceed now to state, and which is proved in A. MARTINEAU [*1*] (p. 149, Theorem 4.1), also in L. HÖRMANDER [*2*] (p. 118, Theorem 2.5.2; Martineau's proof and Hörmander's one are quite different). Let K be a compact subset of $\mathbf{C}^n$; we introduce its "gauge",

$$J_K(\zeta) = \sup_{z \in K} \operatorname{Re}\langle z, \zeta\rangle.$$

Let us say that an analytic functional μ is carried by K if μ is carried by every open neighborhood of K. The announced result can be stated as follows:

Lemma 30.1. – *Let K be a convex compact subset of* $\mathbf{C}^n$, *μ an analytic functional in* C^n. *The following properties are equivalent*:

(a) *μ is carried by K;*

(b) *to every $\varepsilon>0$ there is $C_\varepsilon>0$ such that, for all $\zeta\in\mathbf{C}^n$,*

$$|\hat{\mu}(\zeta)| \leqslant C_\varepsilon \exp(J_K(\zeta) + \varepsilon|\zeta|). \tag{30.4}$$

It is now easy to complete the proof of Theorem 30.2. Indeed, the Fourier-Borel transform of ${}^tP(\partial/\partial x)\,\mu$ is equal to $P\hat{\mu}$. To say that an analytic functional is carried by an open set Ω is equivalent with saying that it is carried by some compact subset of Ω. Suppose that (b), in Lemma (30.1), holds with $\hat{\mu}$ replaced by $P\hat{\mu}$ and for some compact set $K\subset\Omega$. If we apply Lemma 28.1, we see that (30.4) holds (possibly with a modified constant C_ε), whence (a).

Appendix A. Two Lemmas about Fréchet Spaces

We consider two *metric* spaces E and F; they are just topological spaces and need not carry any algebraic structure. We consider also a *continuous* mapping $f: E \to F$.

Lemma A.1. – *Suppose that E is complete and that there is a function of* $t>0$, $\eta(t)>0$, *converging to O with t, such that the following is true*:

$$\text{For every } \varepsilon > O \text{ and every } x \in E, \quad \overline{f(B(x,\varepsilon))} \supset B(f(x), \eta(\varepsilon)). \tag{A 1}$$

Then:

$$\text{For every } \varepsilon \text{ and } \varepsilon' \text{ such that } \varepsilon' > \varepsilon > 0, \text{ and every } x \in E, \quad f(B(x,\varepsilon')) \supset B(f(x), \eta(\varepsilon)). \tag{A 2}$$

(In both spaces E and F we denote by $B(a, \varrho)$ the closed ball centered at a with radius ϱ. In both E and F the metric will be denoted by d.)

Proof of Lemma A.1. – Consider $\varepsilon' > \varepsilon > 0$, $x_0 \in E$ and $y \in B(f(x_0), \eta(\varepsilon))$. Let us select a sequence $\{\alpha_n\}$ $(n=0, 1, \ldots)$ of numbers >0 so as to have

$$\alpha_0 = \varepsilon, \sum_{n=0}^{\infty} \alpha_n = \varepsilon'.$$

There is a point $x_1 \in E$ satisfying

$$d(x_0, x_1) \leqslant \alpha_0 \quad \text{and} \quad d(f(x_1), y) \leqslant \eta(\alpha_1).$$

The existence of x_1 follows from (A.1) which enables us to form, by induction, a sequence of points $x_n \in E$ such that

$$d(x_{n-1}, x_n) \leqslant \alpha_{n-1} \quad \text{and} \quad d(f(x_n), y) \leqslant \eta(\alpha_n).$$

The first property implies that the x_n constitute a Cauchy sequence; as E is complete, it converges to $x \in E$. The second property implies that $f(x_n)$ converges to y, hence $f(x) = y$. Q.E.D.

We consider a metric space T (with no algebraic structure on it) and a Fréchet space E. We are given, for each $t \in T$, a continuous seminorm in E, $\mathfrak{p}_t$. We suppose that the following is true:

For each $x \in E$, the function $t \rightsquigarrow \mathfrak{p}_t(x)$ is continuous in T. (A 3)

Lemma A.2. – *Under the preceding assumptions, the function*

$$(x, t) \rightsquigarrow \mathfrak{p}_t(x) \tag{A 4}$$

is continuous in $E \times T$.

Proof: As the topological space $E \times T$ is metrizable, it suffices to prove the sequential continuity of (A 4). Thus, let us consider a sequence $\{(x_n, t_n)\}$ $(n=1, 2, \ldots)$ in $E \times T$ converging to (x_0, t_0). It follows at once from (A.3) that, for each $x \in E$,

$$\sup_n \mathfrak{p}_{t_n}(x)$$

is finite, hence it is the value at x of some continuous seminorm in E, say $\mathfrak{p}$, in view of Proposition 9.1 and of the fact that E, being a Fréchet space is barrelled (Corollary of Proposition 9.3). Consequently, $\mathfrak{p}(x_n - x_0)$ converges to 0 as $n \to +\infty$.

It suffices then to observe that

$$|\mathfrak{p}_{t_n}(x_n) - \mathfrak{p}_{t_0}(x_0)| \leqslant |\mathfrak{p}_{t_n}(x_0) - \mathfrak{p}_{t_0}(x_0)| + |\mathfrak{p}_{t_n}(x_n) - \mathfrak{p}_{t_n}(x_0)| .$$

The first term in the right hand side converges to zero because of (A 3). The second one is $\leqslant \mathfrak{p}_{t_n}(x_n - x_0) \leqslant \mathfrak{p}(x_n - x_0)$, hence converges also to 0 as $n \to \infty$.

Corollary – *If a bilinear functional f on the product of a Fréchet space E with a metrizable space F is separately continuous, it is continuous.*

We apply Lemma A.2 with $T = F$ and $\mathfrak{p}_t(x) = |f(x, t)|$ $(t \in F)$. We derive from Lemma A.2 that f is continuous at every point (x_0, t_0) such that one at least of the two components x_0 or t_0 is equal to 0. As we have, for arbitrary (x, t) and (x_0, t_0),

$$f(x, t) - f(x_0, t_0) = f(x - x_0, t) + f(x_0, t - t_0),$$

we conclude that f is continuous.

Appendix B. Normal Hilbert Spaces of Distributions

Let Ω be an open subset of an Euclidean space $\mathbf{R}^n$.

Definition B.1. – *A space of distributions in Ω is a vector subspace of $\mathscr{D}'(\Omega)$ equipped with a locally convex topology finer than the one induced by $\mathscr{D}'(\Omega)$.*

Definition B.2. – *A space of distributions in Ω, $\mathscr{M}(\Omega)$, is said to be normal if $\mathscr{C}_c^\infty(\Omega)$ is contained, continuously embedded and dense in $\mathscr{M}(\Omega)$.*

Most spaces of distributions are normal: this is true of $\mathscr{D}'(\Omega)$, $\mathscr{E}'(\Omega)$, $\mathscr{C}^\infty(\Omega)$, etc.; it is also true of the spaces $L^\alpha(\Omega)$ for $1 \leqslant \alpha < +\infty$. But $L^\infty(\Omega)$ is never normal (unless Ω is empty).

If $\mathscr{M}(\Omega)$ is a normal space of distributions in Ω, by transposing the natural injections

$$\mathscr{C}_c^\infty(\Omega) \to \mathscr{M}(\Omega) \to \mathscr{D}'(\Omega),$$

we obtain continuous injections

$$\mathscr{C}_c^\infty(\Omega) \to \mathscr{M}'(\Omega) \to \mathscr{D}'(\Omega). \tag{B.1}$$

In particular, $\mathscr{M}'(\Omega)$ can be regarded as a space of distributions in Ω, not necessarily normal as shown by the choice $\mathscr{M} = L^1$; then $\mathscr{M}' = L^\infty$.

A Hilbert space of distributions in Ω is a space of distributions in Ω whose topology is defined by a Hilbert space structure. We consider two such spaces, $\mathscr{H}_1$ and $\mathscr{H}_2$, and denote by $|||\ \ |||_1$ and $|||\ \ |||_2$ their respective norms. We consider a differential operator in Ω, $\mathbf{P}$.

Lemma B.1. – *Suppose that the Hilbert spaces of distributions in Ω, $\mathscr{H}_1$ and $\mathscr{H}_2$, are normal, and that we have, for some constant $C > 0$ and all functions $\varphi \in \mathscr{C}_c^\infty(\Omega)$,*

$$|||\varphi|||_1 \leqslant C\, |||{}^t\mathbf{P}\varphi|||_2 . \tag{B.2}$$

Then there is a continuous linear map $G: \mathscr{H}'_1 \to \mathscr{H}'_2$, with norm $\leqslant C$, such that, for all $v \in \mathscr{H}'_1$,

$$\mathbf{P}(Gv) = v .$$

Proof: Let M be the vector subspace of $\mathscr{H}_2$ consisting of the elements of the form ${}^t\mathbf{P}\varphi$ with $\varphi \in \mathscr{C}_c^\infty(\Omega)$. Estimate (B.2) implies that the mapping

${}^t\mathbf{P}\varphi \rightsquigarrow \varphi$ is well defined, and continuous from M into $\mathscr{H}_1$ (M carries the norm induced by $\mathscr{H}_2$). By continuity, it can be extended as a continuous linear map of the closure of M in $\mathscr{H}_2$ and then, by defining it as equal to zero on the orthogonal of M, as a continuous linear map G' of $\mathscr{H}_2$ into $\mathscr{H}_1$. The mapping $G: \mathscr{H}'_1 \to \mathscr{H}'_2$ which is the transpose of G' satisfies the requirements of Lemma B.1. Indeed, we have, for all $\varphi \in \mathscr{C}_c^\infty(\Omega)$,

$$\langle \mathbf{P}(Gv), \varphi \rangle = \langle v, G'({}^t\mathbf{P}\varphi) \rangle = \langle v, \varphi \rangle,$$

whence our assertion in view of the density of $\mathscr{C}_c^\infty(\Omega)$ in $\mathscr{H}_1$.

Appendix C. On the Nonexistence of Continuous Right Inverses

Let E, F be two locally convex TVS, $u: E \to F$ a continuous linear map. A continuous linear map $v: F \to E$ is called a right inverse of u if $u \circ v$ is the identity mapping of F.

Lemma C.1. – *The following properties are equivalent:*

(a) *u has a continuous right inverse;*

(b) *u is an epimorphism and* $\operatorname{Ker} u$ *has a topological supplement.*

To say that a subspace M of E has a topological supplement is to say that there is a vector subspace N of E such that the map $(x, y) \rightsquigarrow x+y$ of $M \times N$ into E is an isomorphism. The proof of Lemma C.1 is evident.

Corollary 1. – *Suppose that E and F are Fréchet spaces and that u is surjective. Then u has a continuous right inverse if and only if* $\operatorname{Ker} u$ *has a topological supplement.*

Follows from Lemma C.1 and from the epimorphism theorem (Theorem 10.1).

Theorem C.1. – *Let X be a real analytic manifold, connected and non-compact, of dimension $\geqslant 2$. Let $\mathbf{P}$ be a differential operator in X, having the following two properties:*

for every open subset Ω of X and every function $u \in \mathscr{C}^\infty(\Omega)$, $\mathbf{P}u = 0$ in Ω implies that u is a real analytic function in Ω; (C.1)

every open set $\Omega \subset X$ contains an open subset U such that ${}^t\mathbf{P}u = 0$ for some nonzero distribution u in U. (C.2)

Under these conditions, $\mathbf{P}: \mathscr{C}^\infty(X) \to \mathscr{C}^\infty(X)$ has no continuous right inverse.

If the student is not familiar with manifolds, he may assume that X is an open subset of a Euclidean space $\mathbf{R}^n$. Theorem C.1 is due to A. Grothendieck.

Proof: Let us assume that $\mathbf{P}$ has a continuous right inverse G, and show, under Hypotheses (C.1) and (C.2), that this leads to a contradiction. Let K be a compact subset of X, having a nonempty interior Ω. As G is continuous, there is a continuous seminorm $\mathfrak{p}$ such that, for every $\varphi \in \mathscr{C}^\infty(X)$,

$$\sup_{x \in K} |G\varphi| \leqslant \mathfrak{p}(\varphi).$$

There is a compact set $H \subset X$ such that, if $\operatorname{supp}\varphi \subset X \smallsetminus H$, then $\mathfrak{p}(\varphi)=0$. As X is not compact, we may find an open set $U \subset X$ which does not intersect $H \cup K$. By shrinking U, if necessary, we may assume, in view of (C.2), that ${}^t\mathbf{P}: \mathscr{D}'(U) \to \mathscr{D}'(U)$ is not injective. We may also assume, as X is connected and $\dim X > 1$, that U is the union of an increasing sequence of relatively compact open subsets U_j $(j=1, 2, \ldots)$ such that $X \smallsetminus \overline{U_j}$ is connected.

Let then $\varphi \in \mathscr{C}_c^\infty(U)$; then $\operatorname{supp}\varphi \subset U_j$ for j large enough. We have $P(G\varphi)=0$ in $X \smallsetminus \overline{U_j}$, therefore, by (C.1), $G\varphi$ is an analytic function in $X \smallsetminus \overline{U_j}$. But on the other hand, $\varphi=0$ in some neighborhood of H, hence by (C.3), $G\varphi=0$ in K, a fortiori in $\mathcal{O}$, which implies at once that $G\varphi=0$ in $X \smallsetminus \overline{U_j}$. In other words, $G\varphi \in \mathscr{C}_c^\infty(U)$. This implies that the restriction of $\mathbf{P}\mathscr{C}_c^\infty(U)$ maps this space *onto* itself, contradicting the fact that its transpose ${}^t\mathbf{P}: \mathscr{D}'(U) \to \mathscr{D}'(U)$ is not injective. Q.E.D.

The elliptic operators satisfy both conditions (C.1) and (C.2) provided that their coefficients be analytic (cf Example 26.1).

There are differential operators which satisfy both (C 1) and (C 2) and which are not elliptic, e.g., the operators in $\mathbf{R}^2$, $(\partial/\partial x_1) + i x_1^{2k} (\partial/\partial x_2)$ $(k=1, 2, \ldots$; cf. Example 26.2 and MIZOHATA [*1*]).

Main Definitions and Results
Concerning the Spectrum of a Locally Convex Space

Let E be a vector space (either complex or real). The *total spectrum* of E, SPECE, is the cone of all the seminorms on E, with its natural order relation: $\mathfrak{p} \leqslant \mathfrak{q}$ means $\mathfrak{p}(x) \leqslant \mathfrak{q}(x)$ for all $x \in E$. An *irreducible* subset A of SPECE is a subcone of E with the following properties: i) if $\mathfrak{p}$, $\mathfrak{q} \in A$, $\sup(\mathfrak{p}, \mathfrak{q}) \in A$; ii) if $\mathfrak{p} \in A$ and $\mathfrak{q} \in \operatorname{Spec} E$, $\mathfrak{q} \leqslant \mathfrak{p}$, then $\mathfrak{q} \in A$. The datum of a *locally convex topology* on E is equivalent with the datum of an irreducible subset of SPECE, precisely the set of all the seminorms which are continuous in the given topology. If E is a locally convex space, the set of the continuous seminorms on E is the *spectrum* of E, denoted by Spec$E$.

Lemma 1.1. – *Let $M \subset E$ be a vector subspace, $\mathfrak{p} \in$ SPECM, $\mathfrak{q} \in$ SPECE be such that the restriction of $\mathfrak{q}$ to M is $\geqslant \mathfrak{p}$. There exists $\bar{\mathfrak{p}} \in$ SPECE, $\bar{\mathfrak{p}} \leqslant \mathfrak{q}$, $\bar{\mathfrak{p}}$ extending $\mathfrak{p}$.*

Lemma 3.1. – *Let E be a locally convex space, A a closed convex balanced subset of E, K a compact subset of E which does not intersect A. There is $\mathfrak{p} \in \operatorname{Spec} E$, $\mathfrak{p} \leqslant 1$ on A, $\mathfrak{p} > 1$ on K.*

A set A is *balanced* if $\lambda x \in A$ for all $x \in A$ and all scalars such that $|\lambda| \leqslant 1$.

Let E and F be two locally convex spaces (over the same scalar field: either **R** or **C**), $u: E \to F$ a continuous linear map. We set, for $\mathfrak{q} \in \operatorname{Spec} F$, $u_* \mathfrak{q} = \mathfrak{q} \circ u$ ($\circ$: compose); $u_*: \operatorname{Spec} F \to \operatorname{Spec} E$ is a morphism for the structures of ordered cones.

Proposition 4.2. – Imu *is dense if and only if u_* is injective.*

Proposition 4.3. – *If u is a monomorphism, u_* is surjective; the converse is true when E and F are Hausdorff* (Proposition 4.6).

Corollary of Proposition 4.6: *If E and F are Hausdorff and complete, u is an isomorphism if and only if u_* is bijective.*

For $A \subset E$, we set $A^0 = \{\mathfrak{p} \in \operatorname{Spec} E; \sup \mathfrak{p}(x) \leqslant 1\}$; A^0 is the *polar* of A.

Proposition 4.7. – *If F is Hausdorff, u is a homomorphism if and only if* $\operatorname{Im} u^* = (\operatorname{Ker} u)^0$.

Let $\mathfrak{p} \in \operatorname{Spec} E$, and let $E_{(\mathfrak{p})}$ be the vector space E equipped with the topology defined by the single seminorm $\mathfrak{p}$. Let $E_\mathfrak{p} = E_{(\mathfrak{p})}/\operatorname{Ker} \mathfrak{p}$ be the associated normed space, and $\hat{E}_\mathfrak{p}$ its completion: $\hat{E}_\mathfrak{p}$ is a Banach space, and the compose of the natural mappings $E \to E_{(\mathfrak{p})} \to E_{\mathfrak{p}} \to \hat{E}_{\mathfrak{p}}$ defines the natural mapping $\omega_{\mathfrak{p}}$ of E into $\hat{E}_{\mathfrak{p}}$. We look at the collection of Banach spaces $\hat{E}_{\mathfrak{p}}$, $\mathfrak{p} \in \operatorname{Spec} E$, as a *fiber set* over $\operatorname{Spec} E$. If $\mathfrak{p}, \mathfrak{q} \in \operatorname{Spec} E$ and $\mathfrak{q} \leqslant \mathfrak{p}$, the identity mapping of E is continuous from $E_{(\mathfrak{p})}$ onto $E_{(\mathfrak{q})}$ and induces a continuous linear map $\omega_{\mathfrak{q}}^{\mathfrak{p}}: \hat{E}_{\mathfrak{p}} \to \hat{E}_{\mathfrak{q}}$ (with norm $\leqslant 1$). Let S be a subset of $\operatorname{Spec} E$; a *regular section* over S is a mapping s such that $s(\mathfrak{p}) \in \hat{E}_{\mathfrak{p}}$ for every $\mathfrak{p} \in S$, and such that $\omega_{\mathfrak{q}}^{\mathfrak{p}}(s(\mathfrak{p})) = s(\mathfrak{q})$ for all $\mathfrak{p}, \mathfrak{q} \in S$ verifying $\mathfrak{q} \leqslant \mathfrak{p}$. If $x \in E$, $\mathfrak{p} \rightsquigarrow \omega_{\mathfrak{p}}(x)$ is a regular section over any subset S of $\operatorname{Spec} E$, denoted by $\omega_{\mathfrak{q}}^{\mathfrak{p}}(x)$. Let $\Gamma(S)$ be the set of all regular sections over S: it has a natural structure of vector space, also a natural locally convex topology: the topology of pointwise convergence in S. If S is irreducible, $x \rightsquigarrow \omega_S(x)$ is a continuous linear map, with dense image, of E into $\Gamma(S)$. If E is Hausdorff and complete, and if $S = \operatorname{Spec} E$, this mapping is an isomorphism of E onto $\Gamma(\operatorname{Spec} E)$.

Let now F be another locally convex space, $u: E \to F$ a continuous linear map. Let $\mathfrak{p} \in \operatorname{Spec} E$, $\mathfrak{q} \in \operatorname{Spec} F$ be such that $u_* \mathfrak{q} \leqslant \mathfrak{p}$; then u induces a continuous linear map (with norm $\leqslant 1$) of $\hat{E}_{\mathfrak{p}}$ into $\hat{F}_{\mathfrak{q}}$, denoted by $u_{\mathfrak{q}}^{\mathfrak{p}}$; if $\mathfrak{p} = u_* \mathfrak{q}$, we write $u_{\mathfrak{q}}$ instead of $u_{\mathfrak{p}}^{u_* \mathfrak{q}}$; $u_{\mathfrak{q}}$ is an isometry of $\hat{E}_{u_* \mathfrak{q}}$ into $\hat{F}_{\mathfrak{q}}$. Let $A \in \operatorname{Spec} E$, $B = \bar{u}_*^{1}(A)$, s be a regular section over A; the *direct image of s under the mapping u, us*, is the regular section over B defined by

$$us(\mathfrak{q}) = u_{\mathfrak{q}}(s(u_* \mathfrak{q})), \qquad \mathfrak{q} \in B.$$

An equicontinuous subset of $\operatorname{Spec} E$ is a subset A such that there is $\mathfrak{p}_0 \in \operatorname{Spec} E$ such that $\mathfrak{p} \leqslant \mathfrak{p}_0$ for all $\mathfrak{p} \in A$.

Definition 10.1. – *u is called presurjective if, for every equicontinuous subset A of* $\operatorname{Spec} E$ *and for every $y \in F$, there is a regular section s over A such that $us = \varpi_B(y)$, $B = \bar{u}_*^{1}(A)$.*

Theorem 10.1 – ("the epimorphism theorem"). – *If E and F are Fréchet spaces, the following properties are equivalent:*

(1) *u is an epimorphism;*

(2) *u is surjective;*

(3) *u is a homomorphism with dense image;*

(4) *u_* is injective and* $\operatorname{Im} u_* = (\operatorname{Ker} u)^0$;

(5) *if u is a neighborhood of* 0 *in E, $\overline{u(U)}$ is a neighborhood of* 0 *in F;*

(6) *the preimage under u_* of any equicontinuous subset of* $\operatorname{Spec} E$ *is an equicontinuous subset of* $\operatorname{Spec} F$;

(7) *u is presurjective.*

If we do not assume that E and F are Fréchet spaces, we have the net of implications of p. 32. The closed graph theorem for Fréchet spaces (Theorem 12.1) follows from the equivalence (1)⇔(2) in standard manner.

Let $\mathfrak{p} \in \mathrm{SPEC}\,E$; we denote by $S_{\mathfrak{p}}$ the set of $\mathfrak{q} \in \mathrm{SPEC}\,E$ such that $\mathfrak{q} \leqslant C\mathfrak{p}$ for some constant $C>0$; the infimum of these constants C is denoted by $\|\mathfrak{q}\|_{\mathfrak{p}}$. Let $\mathscr{B}$ (resp. $\mathscr{B}'$) be a basis of continuous seminorms (Definition 2.3) in E (resp. F); consider the relation:

$$\begin{aligned} &\forall \mathfrak{p} \in \mathscr{B}, \exists \mathfrak{q} \in \mathscr{B}' \text{ and } C>0 \text{ such that, for all } \mathfrak{r} \in \mathrm{Spec}\,F, \\ &u_* \mathfrak{r} \in S_{\mathfrak{p}} \text{ implies } \mathfrak{r} \in S_{\mathfrak{q}} \text{ and } \|\mathfrak{r}\|_{\mathfrak{q}} \leqslant C \|u_* \mathfrak{r}\|_{\mathfrak{p}}. \end{aligned} \tag{11.1}$$

Proposition 11.1. – *If there are bases $\mathscr{B}$, $\mathscr{B}'$ satisfying* (11.1), *u is presurjective. Conversely, if F is barrelled* (Definition 9.2), *and if u is presurjective,* (11.1) *holds for all pairs of bases $\mathscr{B}$, $\mathscr{B}'$.*

Definition 11.1. – *The pair (E, F) is said to be u-convex if $\forall \mathfrak{p} \in \mathrm{Spec}\,E$, $\exists \mathfrak{q} \in \mathrm{Spec}\,F$ such that, for all $\mathfrak{r} \in \mathrm{Spec}\,F$, $u_* \mathfrak{r} \leqslant \mathfrak{p}$ implies* $\mathrm{Ker}\,\mathfrak{q} \subset \mathrm{Ker}\,\mathfrak{r}$.

Definition 11.2. – *The mapping u is said to be semiglobally surjective if $\forall \mathfrak{q} \in \mathrm{Spec}\,F$, $\forall y \in F$, $\exists x \in E$ such that $u(x) - y \in \mathrm{Ker}\,\mathfrak{q}$.*

Proposition 11.2. – *If (E, F) is u-convex and u is semiglobally surjective, then u is presurjective.*

Consider four locally convex Hausdorff spaces, E_0, F_0, E, F, and three continuous linear mappings

$$u_0 : E_0 \to F_0, \quad \varrho : E_0 \to E, \quad \sigma : F_0 \to F.$$

Definition 13.1. – *u_0 is said to be essentially univalent with respect to (ϱ, σ) if $\forall \mathscr{F}$, filter on E_0, if $\varrho(\mathscr{F}) \to 0$ in E and if $\sigma(u_0(\mathscr{F}))$ converges in F, we must have* $\lim \sigma(u_0(\mathscr{F})) = 0$.

Let G be the closure in $E \times F$ of the set of pairs $(\varrho(x_0), \sigma(u_0(x_0)))$ with $x_0 \in E_0$: *the fact that u_0 is essentially univalent with respect to (ϱ, σ) is equivalent with the fact that the first coordinate projection, $(x, y) \rightsquigarrow x$, is injective on G.* The image of G under $(x, y) \rightsquigarrow x$ will then be denoted by $\tilde{E}_0$ and the mapping $\tilde{u}_0 : \tilde{E}_0 \to F$ which is the compose $x \rightsquigarrow (x, y) \rightsquigarrow y$ will be called the *canonical extension* of u_0 with respect to (ϱ, σ).

Definition 14.1. – *We say that u_0 has the homogeneous approximation property if $\varrho(\mathrm{Ker}\,u_0)$ is dense in $\mathrm{Ker}\,\tilde{u}_0$.*

Theorem 14.1 ("existence and approximation of solutions"). – *Suppose that u_0 is essentially univalent with respect to (ϱ, σ). The following properties are equivalent:*

(a) *the canonical extension $\tilde{u}_0$ of u_0 is a homomorphism and has the homogeneous approximation property;*

(b) $(\mathrm{Im}\,\varrho_*) \cup (\mathrm{Ker}\,u_0)^0 \subset u_{0*}(\mathrm{Im}\,\sigma_*)$.

We come now to the relation with duality.

Lemma 15.1 (Hahn-Banach theorem). – *Let $M \subset E$ be a vector subspace, $\mathfrak{p} \in \mathrm{SPEC}\,E$, x_0^* a linear functional on M such that $|\langle x^*, x\rangle| \leqslant \mathfrak{p}(x)$ for all $x \in M$. There is a linear functional x^* on E, extending x_0^* and such that $|x^*| \leqslant \mathfrak{p}$* (i.e., $\forall x \in E$, $|\langle x^*, x\rangle| \leqslant \mathfrak{p}(x)$).

Corollary 1. – *Given $\mathfrak{p} \in \mathrm{SPEC}\,E$, $x \in E$ arbitrary, there is $x^* \in E^*$ such that $\langle x^*, x\rangle = \mathfrak{p}(x)$ and $|x^*| \leqslant \mathfrak{p}$.*

Corollary 2. – *Any seminorm $\mathfrak{p}$ on E is the upper enveloppe of the seminorms $|x^*|$, $x^* \in E^*$, which are $\leqslant \mathfrak{p}$.*

We have denoted by E^* the *algebraic dual* of E, that is, the vector space of all linear functionals on E.

Lemma 15.2 (Mackey's theorem). – *A seminorm $\mathfrak{p} \in \mathrm{SPEC}\,E$ is bounded on a subset A of E if and only if every seminorm $|x^*| \leqslant \mathfrak{p}$, $x^* \in E^*$, is bounded on A.*

The seminorms $\mathfrak{p}$ of the kind $\mathfrak{p} = |x^*|$, $x^* \in E^*$, are characterized, among all the seminorms on E, by the fact that $\hat{E}_{\mathfrak{p}}$ is a one-dimensional Hilbert space, or else by the fact that they are *extremal*, in the sense that if $\mathfrak{q} \in \mathrm{SPEC}\,E$, $\mathfrak{q} \leqslant |x^*|$, then we must have $\mathfrak{q} = |\lambda x^*|$, λ a scalar.

Let F be another vector space, $u: E \to F$ a linear map. Let us suppose from now on that E and F carry locally convex Hausdorff topologies and that u is continuous; then E' (resp. F') denotes the dual, i.e. the vector space of *continuous* linear functionals, of E (resp. of F), ${}^t u: F' \to E'$ the transpose of u.

Proposition 16.1. – *$x' \in E'$ belongs to $\mathrm{Im}\,{}^t u$ if and only if the continuous seminorm $|x'|$ belongs to $\mathrm{Im}\,u^*$.*

Proposition 16.3. – *Suppose that*

$$\overline{\mathrm{Im}\,u}, \text{ closure of } \mathrm{Im}\,u \text{ in } F, \text{ is a barrelled space (Definition 9.2);} \tag{16.2}$$

$$\forall y \in \overline{\mathrm{Im}\,u} \text{ belongs to the closure of a subset of } \mathrm{Im}\,u \text{ which is bounded in } F. \tag{16.3}$$

Then $\mathfrak{p} \in \mathrm{Spec}\,E$ belongs to $\mathrm{Im}\,u^*$ if and only if every $|x'| \leqslant \mathfrak{p}$, $x' \in E'$, belongs to $\mathrm{Im}\,u^*$.

Properties (16.2) and (16.3) are trivially satisfied when F is a Fréchet space; (16.2) is valid whenever F is a barrelled space and $\mathrm{Im}\,u$ is dense in it.

Corollaries 2, 3, 5. 4, – *Suppose that* (16.2) *and* (16.3) *hold. Then*:

2) *u is a homomorphism if and only if $\mathrm{Im}\,{}^t u = (\mathrm{Ker}\,u)^\perp$;*
3) *u is a homomorphism with dense image if and only if ${}^t u$ is injective and $\mathrm{Im}\,{}^t u = (\mathrm{Ker}\,u)^\perp$;*
4) *u is a monomorphism if and only if ${}^t u$ is surjective;*
5) *u is an isomorphism if and only if ${}^t u$ is bijective.*

We recall that if $M \subset E$, $M^{\perp} = \{x' \in E'; x' = 0 \text{ on } M\}$. Theorem 10.1 yields:

Theorem 17.1. – *Suppose that E and F are Fréchet spaces; the following properties are equivalent:*

(a)–(b)–(c)–(d): *u is an epimorphism – surjective – a homomorphism with dense image – presurjective;*

(e) *the preimage under ${}^t u$ of any equicontinuous subset of E' is an equicontinuous subset of F';*

(f) *${}^t u$ is injective and* $\mathrm{Im}\,{}^t u = (\mathrm{Ker}\, u)^{\perp}$.

Proposition 17.1. – *Suppose there are bases of continuous seminorms $\mathscr{B}$, $\mathscr{B}'$ in E and F respectively, such that the following holds:*

$$\begin{array}{l} \forall \mathfrak{p} \in \mathscr{B},\ \exists \mathfrak{q} \in \mathscr{B}',\ \exists C > 0 \text{ such that, for all } y' \in F', \\ {}^t u(y') \in (\hat{E}_{\mathfrak{p}})' \text{ implies } y' \in (\hat{F}_{\mathfrak{q}})' \text{ and} \end{array} \tag{17.1}$$

$$\|y'\|_{\mathfrak{q}} \leqslant C \|{}^t u(y')\|_{\mathfrak{p}}.$$

Then u is presurjective. Conversely, if F is barrelled, (17.1) *holds for all pairs of bases $\mathscr{B}$, $\mathscr{B}'$ of continuous seminorms.*

Proposition 17.1 gives the reason why "a priori estimates" are so important in the theory of existence of solutions to linear PDEs.

Proposition 17.2. – *(E, F) is u-convex* (Definition 11.1) *if and only if, $\forall \mathfrak{p} \in \mathrm{Spec}\, E$, $\exists \mathfrak{q} \in \mathrm{Spec}\, F$ such that, for all $y' \in F'$,*

$${}^t u(y') \in (\hat{E}_{\mathfrak{p}})' \quad \textit{implies} \quad y' \in (\mathrm{Ker}\, \mathfrak{q})^{\perp}.$$

Theorem 14.1 yields:

Theorem 17.2, Corollary 2. – *Same situation as in Theorem* 14.1. *Suppose furthermore that the following holds:*

$$\overline{\sigma(\mathrm{Im}\, u_0)} \subset F \text{ is barrelled;} \tag{17.3}$$

$$\begin{array}{l} \forall y \in \overline{\sigma(\mathrm{Im}\, u_0)} \text{ belongs to the closure of a subset of} \\ \sigma(\mathrm{Im}\, u_0) \text{ which is bounded in } F; \end{array} \tag{17.4}$$

$$\mathrm{Im}\,{}^t u_0 = (\mathrm{Ker}\, u_0)^{\perp}; \tag{17.5}$$

$${}^t u_0 \text{ is injective (i.e., } \mathrm{Im}\, u_0 \text{ is dense in } F_0). \tag{17.6}$$

The following two facts are equivalent:

(a) *the canonical extension $\tilde{u}_0$ of u_0 is a homomorphism and has the homogeneous approximation property;*

(b) *for every $y_0' \in F_0'$,*

$${}^t u_0(y_0') \in \mathrm{Im}\,{}^t \varrho \quad \textit{implies} \quad y_0' \in \mathrm{Im}\,{}^t \sigma.$$

Corollary 3. – *We assume that E_0, F_0, E, F are Fréchet spaces and that u_0 is essentially univalent with respect to (ϱ, σ). If u_0 is surjective, the following facts are equivalent:*

(a) *the canonical extension of* u_0 *is an epimorphism and has the homogeneous approximation property;*

(b) $\sigma(F_0)$ *is dense in* F *and for every* $y_0' \in F$,

$$ {}^t u_0(y_0') \in \operatorname{Im}{}^t\varrho \quad \textit{implies} \quad y_0' \in \operatorname{Im}{}^t\sigma . $$

We recall that the canonical extension $\tilde{u}_0$ of u_0 is a continuous linear map of a subspace $\tilde{E}_0$ of E into F ($\tilde{E}_0$ is the "completion" of E for the "graph topology" – whenever E and F are complete.)

Some Definitions in PDE Theory

Analytic hypoelliptic operators. See Definition 26.3.

Characteristic. Let **P** be a differential operator on a manifold Ω, $P_m(x, i\xi)$ its principal symbol (see below). A vector ξ of the cotangent space to Ω at the point x is said to be characteristic if $P_m(x, i\xi)=0$. A submanifold S of codimension 1 of Ω is said to be characteristic with respect to **P** at some point $x \in S$ if the normal to S at x is characteristic.*)

Cauchy problem. Let Ω, S, **P** be like above. Suppose that **P** is of order $m \geqslant 1$ in Ω, and let $(\partial/\partial\nu)$ denote the differentiation of functions along the normal to S. The Cauchy problem is the problem of finding a function u in Ω, say of class $\mathscr{C}^m$, verifying

$$\mathbf{P}u = f \text{ in } \Omega\,, \quad (\partial/\partial\nu)^k u = g_k \quad \text{on } S \text{ for } k = 0, \ldots, m-1\,,$$

where f and the $g_k (0 \leqslant k \leqslant m)$ are given functions, the first one defined in Ω, the latter, called *Cauchy data*, on S.*)

Dirichlet problem. – Here Ω must be a manifold with boundary, and **P** an operator of order *two*. The Dirichlet problem is the problem of finding a function u in Ω, verifying

$$\mathbf{P}u = f \text{ in } \Omega\,, \quad u = g \text{ on } \partial\Omega, \text{ boundary of } \Omega\,,$$

where f and g are given functions, on Ω and on $\partial\Omega$ respectively.

Elliptic operators. 1) *Scalar elliptic operators.* Let **P** be a differential operator on some manifold Ω, $P^0(x, \xi)$ its principal symbol (see below). The operator **P** is said to be elliptic in Ω if, for any $x \in \Omega$, the (homogeneous) polynomial with respect to ξ, $P^0(x, \xi)$, does not vanish in $\mathbf{R}^n \smallsetminus \{0\}$. *Examples of elliptic differential operators*: the Laplace operator in $\mathbf{R}^n$, more generally the Laplace-Beltrami operator on a Riemannian manifold; also, the Cauchy-Riemann operator $\partial/\partial\bar{z}$ in $\mathbf{R}^2$.

2) *Elliptic systems.* Various definitions are used, depending on the problem at hand. The following one is certainly among the most general. Let

---

*) There is an analog definition in the complex domain: Ω, S must then be complex analytic manifolds and all intervening functions are holomorphic.

$\mathbf{P}$ be a $N_1 \times N_2$ differential system, defined in Ω. In other words, $\mathbf{P}$ is a matrix with N_1 columns and N_2 rows, whose entries are scalar differential operators $\mathbf{P}_{jk}$ in Ω; $\mathbf{P}$ acts on functions whose values are vectors with N_1 components, $\mathbf{f}=(f_1,\ldots,f_{N_1})$, in the manner

$$(\mathbf{Pf})_j = \sum_{j=1}^{N_1} \mathbf{P}_{jk} f_j, \qquad k = 1, \ldots, N_2 .$$

Suppose then that there exist N_1 integers s_j and N_2 integers t_k such that, for each pair (j, k), the differential operator $\mathbf{P}_{jk}$ is exactly of order $s_j - t_k$ (hence all these numbers must be $\geqslant 0$). Let then $P_{jk}^0(x, \xi)$ be the principal symbol of $\mathbf{P}_{jk}$, and consider the matrix $(P_{jk}^0(x, \xi))$. It defines a linear map of $\mathbf{C}^{N_1}$ into $\mathbf{C}^{N_2}$; suppose that this map is *injective* for all $x \in \Omega$ and all $\xi \in \mathbf{R}^n$, $\xi \neq 0$ (which requires $N_1 \leqslant N_2$). Under these assumptions, $\mathbf{P}$ is said to be elliptic. *Examples of elliptic systems*: the gradient; any diagonal system with diagonal entries which are elliptic scalar operators.

By using local charts, one defines easily *elliptic $V-W$ differential operators*, where V and W are vector fibre bundles over the same base manifold. For example, the *exterior differential d*, acting from functions into sections of the cotangent bundle, i.e., into 1-forms, is elliptic.

Elliptic systems with analytic coefficients are analytic hypoelliptic (Definition 26.3). As a matter of fact, the scalar elliptic operators with constant coefficients are the only analytic hypoelliptic differential operators with constant coefficients (result due to PETROWSKY; see e.g. TREVES [*1*], Th. 7.8). Determined (i.e., for which $N_1 = N_2$) elliptic systems are the differential operators with the best properties concerning boundary value problems (see HÖRMANDER [*1*], Ch. X) – just as the Laplace operator is well suited for the study of the "first boundary value problem" (or *Dirichlet problem*, see above).

Fundamental kernel, fundamental solution. Let $\mathbf{P}$ be a differential operator on some manifold Ω. A fundamental kernel of $\mathbf{P}$ is a kernel-distribution $K(x, y)$ on $\Omega_x \times \Omega_y$ such that, for every function $u \in \mathscr{C}_c^\infty(\Omega)$,

$$u(x) = \mathbf{P}_x \int K(x, y)\, u(y)\, dy = \int K(x, y)\, \mathbf{P}u(y)\, dy$$

(we have denoted by $\int T(y)\, v(y)\, dy$ the value of the distribution T on the test function v; what we call here distributions are continuous linear functionals over the space of test-functions). What we have just defined is sometimes called a *two-sided* fundamental kernel of $\mathbf{P}$; one-sided fundamental kernels are easy to define (either by $\mathbf{P}_x K(x, y) = \delta(x-y)$ or by ${}^t\mathbf{P}_y K(x, y) = \delta(x-y)$).

The notion of fundamental solution applies to differential operators with constant coefficients. If $P(\partial/\partial x)$ is such an operator, say on $\mathbf{R}^n$, a fundamental solution of $P(\partial/\partial x)$ is a distribution E on $\mathbf{R}^n$ such that $P(\partial/\partial x)\, E = \delta$, Dirac's measure. The kernel $E(x-y)$ is then a (two-sided) fundamental kernel of $P(\partial/\partial x)$. It is a result due to EHRENPREIS and MALGRANGE that every nonidentically zero differential polynomial has a fundamental solution. (see TREVES [*1*], Th. 3.1).

Hyperbolic operators. See **Strongly hyperbolic** operators below.

Hypoelliptic operators. See Definition 21.2.

Parabolic operators. – 1) *Second order parabolic operators.* Let $\mathbf{P}$ be a second order differential operator on a manifold Ω, and let $P^0(x, \xi)$ be its principal symbol (see below). The operator $\mathbf{P}$ is said to be *parabolic* if $P^0(x, \xi)$ is real for $x \in \Omega$, and $\xi \in \mathbf{R}^n$, does not change sign when x and ξ vary, and if, for every $x \in \Omega$, the set of real zeros $\xi \in \mathbf{R}^n$ of the polynomial with respect to ξ, $P^0(x, \xi)$, is at most one dimensional (if it is zero dimensional, $\mathbf{P}$ is elliptic). *Example of a nonelliptic parabolic differential operator*: the operator in $\mathbf{R}^2$, $(\partial/\partial x_1)^2 + 2x_1(\partial/\partial x_1)(\partial/\partial x_2) + x^2(\partial/\partial x_2)^2$.

2) *Parabolic operators in the sense of Petrowsky.* Let $P(\partial/\partial x, \partial/\partial t)$ be a differential operator with constant coefficients on $\mathbf{R}^n \times \mathbf{R}^1$; $x = (x_1, \ldots, x_n)$ are the space variables, t the time. Let d, m be integers $\geqslant 1$ such that $P(\xi, \tau^d)$ is of degree md (as a polynomial with respect to (ξ, τ)), and let $P^0(\xi, \tau)$ be the part of $P(\xi, \tau)$ such that $P^0(\xi, \tau^d)$ is the homogeneous part of degree maximum $(= md)$ of $P(\xi, \tau^d)$. The differential operator $P(\partial/\partial x, \partial/\partial t)$ is said to be *d-parabolic* (in the sense of PETROWSKY) if there is a number $\delta > 0$ such that, for all $\xi \in \mathbf{R}^n$, $|\xi| = 1$, $P^0(\xi, \tau) = 0$ implies $|\mathrm{Im}\, \tau| \leqslant -\delta$. In this case, m and d are uniquely determined, and d must be *even. Example of a parabolic operator*: the heat operator $\Delta - (\partial/\partial t)$, where Δ is the Laplace operator in $\mathbf{R}^n$, is *two*-parabolic.

Let now $P(x, t, \partial/\partial x, \partial/\partial t)$ be a differential operator with variable coefficients on $\mathbf{R}^n \times \mathbf{R}^1$. It is said d-parabolic if it is so when one fixes x and t arbitrarily (and if it keeps constant order).

Parabolic operators are hypoelliptic (Definition 21.2). All the functions which satisfy a homogeneous equation $P(\partial/\partial x, \partial/\partial t)\, u = 0$, where $P(\partial/\partial x, \partial/\partial t)$ is d-parabolic, belong to d-th Gevrey class. Parabolic operators have interesting properties in relation with the so-called "mixed problems" (these are evolution problems, with prescribed data at time $t = 0$, and at the boundary of some space domain at each time t).

Principal symbol of a differential operator. Let **P** be a differential operator on a manifold Ω. Let $(x_1, \ldots, x_n)$ be local coordinates, with domain, an open subset U of Ω. In these coordinates, **P** can be expressed by

$$P(x, \partial/\partial x) = \sum_{|p| \leqslant m} a_p(x) (\partial/\partial x)^p,$$

assuming that at least one of the coefficients a_p, with $|p| = m$, does not vanish identically in U. Let

$$P_m(x, \partial/\partial x) = \sum_{|p| = m} a_p(x) (\partial/\partial x)^p$$

be the *principal part* of $P(x, \partial/\partial x)$ (also called the *leading terms*). The principal symbol of **P** is then obtained by substituting $i\xi_j$ for $\partial/\partial x_j$ in the principal part. It is a homogeneous polynomial of degree m with respect to ξ with coefficients in the ring of (complex) functions of x. Changes of (local) coordinates transform the principal part $P_m(x, \partial/\partial x)$ like a homogeneous covariant tensor of degree m. In view of this, the principal symbol $P_m(x, i\xi)$ can be regarded as a function on the cotangent bundle over Ω.

Sobolev spaces. See p. 59.

Strongly hyperbolic operators. Let N be a nonzero vector of $\mathbf{R}^n$, **P** a differential operator in some open subset Ω of $\mathbf{R}^n$, $P_m(x, i\xi)$ its principal symbol (see above). The operator **P** is said to be *strongly hyperbolic in Ω, in the direction N*, if, for all $x \in \Omega$ and all $\xi \in \mathbf{R}^n$, linearly independent of N, the polynomial with respect to τ, $P_m(x, \tau N + \xi)$, has *m distinct real* roots. *Example*: the wave operator in $\mathbf{R}^n$, $(\partial/\partial x_1)^2 + \cdots + (\partial/\partial x_{n-1})^2 - (\partial/\partial x_n)^2$, is strongly hyperbolic in $\mathbf{R}^n$, in the direction $N = (0, \ldots, 0, 1)$.

The strongly hyperbolic operators (also called *hyperbolic in the sense of Petrowsky*) have the best properties in relation with the Cauchy problem (see above).

Bibliographic References

Agranovich, M. S.: [*1*] Partial differential equations with constant coefficients, Uspehi Mat. Nauk, **16**, No. 2, 27 (1961). (Russian; English translation in *Russian Mathematical Surveys*.)

Bourbaki, N.: [*1*] Espaces Vectoriels Topologiques, 2 vol., *Hermann* Paris 1953 & 1955.

Harvey, C. M.: [*1*] On Domination Estimates and Global Existence, *J. of Math. and Mechan.*, **16**, No. 7, 675 (1967).

Hörmander, L.: [*1*] Linear Partial Differential Operators, *Springer* 1963.

[*2*] L_2 Estimates and Existence Theorems for the $\bar{\partial}$ Operator, *Acta Math.* **113,** 89 (1965).

Lions, J. L.: [*1*] Supports dans la transformation de Laplace, *J. Analyse Math.*, **2,** 369 (1952–53).

Malgrange, B.: [*1*] Existence et approximation des solutions des équations aux dérivées partielles et des équations de convolution, *Ann. Inst. Fourier Grenoble*, **6,** 271 (1955–56).

Martineau, A.: [*1*] Sur les fonctionnelles analytiques et la transformation de Fourier-Borel, *J. Analyse Math.*, **9,** 1 (1963).

[*2*] Equations différentielles d'ordre infini, Séminaire Leray, Collège de France, 1965–66.

Mizohata, S.: [*1*] Solutions nulles et solutions non analytiques, *J. Math. Kyoto Univ.*, Vol. 1 (1962), p. 271.

Treves, F.: [*1*] Linear Partial Differential Equations with constant coefficients, *Gordon & Breach* New York 1966.

[*2*] Topological Vector Spaces, Distributions and Kernels, *Academic Press* New York 1967.

[*3*] Cours sur les équations aux dérivées partielles linéaires, Faculté des Sciences, Paris 1965–66.

Printed by N.V. Drukkerij D. Reidel, Dordrecht-Holland